W0206974

THE EUROPEAN UNION SERIES

General Editors: Neill Nugent, William E. Paterson

The European Union series provides an authoritative library on the European Union, ranging from general introductory texts to definitive assessments of key institutions and actors, issues, policies and policy processes, and the role of member states.

Books in the series are written by leading scholars in their fields and reflect the most up-to-date research and debate. Particular attention is paid to accessibility and clear presentation for a wide audience of students, practitioners and interested general readers.

The series editors are **Neill Nugent**, Emeritus Professor of Politics at Manchester Metropolitan University, UK, and **William E. Paterson**, Honorary Professor in German and European Studies, University of Aston. Their co-editor until his death in July 1999, **Vincent Wright**, was a Fellow of Nuffield College, Oxford University.

Feedback on the series and book proposals are always welcome and should be sent to Stephen Wenham, Palgrave, 4 Crinan Street, London N1 9XW, or by e-mail to s.wenham@palgrave.com

---

General textbooks

Published

Laurie Buonanno and Neill Nugent **Policies and Policy Processes of the European Union**
Desmond Dinan **Encyclopedia of the European Union** [Rights: Europe only]
Desmond Dinan **Europe Recast: A History of the European Union (2nd edn)** [Rights: Europe only]
Desmond Dinan **Ever Closer Union: An Introduction to European Integration (4th edn)** [Rights: Europe only]
Mette Eilstrup Sangiovanni (ed.) **Debates on European Integration: A Reader**
Simon Hix and Bjørn Høyland **The Political System of the European Union (3rd edn)**
Dirk Leuffen, Berthold Rittberger and Frank Schimmelfennig **Differentiated Integration**
Paul Magnette **What is the European Union? Nature and Prospects**
John McCormick **Understanding the European Union: A Concise Introduction (6th edn)**
Brent F. Nelsen and Alexander Stubb **The European Union: Readings on the Theory and Practice of European Integration (4th edn)** [Rights: Europe only]
Neill Nugent (ed.) **European Union Enlargement**
Neill Nugent **The Government and Politics of the European Union (7th edn)**
John Peterson and Elizabeth Bomberg **Decision-Making in the European Union**
Ben Rosamond **Theories of European Integration**
Sabine Saurugger **Theoretical Approaches to European Integration**
Ingeborg Tömmel **The European Union: What it is and How it Works**
Esther Versluis, Mendeltje van Keulen and Paul Stephenson **Analyzing the European Union Policy Process**
Hubert Zimmermann and Andreas Dür (eds) **Key Controversies in European Integration**

Forthcoming

Magnus Ryner and Alan Cafruny **A Critical Introduction to the European Union**

Also planned

The European Union and Global Politics
The Political Economy of European Integration

Series Standing Order (outside North America only)
ISBN 978-0-333-71695-3 hardback
ISBN 978-0-333-69352-0 paperback
Full details from www.palgrave.com

The major institutions and actors

Published

Renaud Dehousse **The European Court of Justice**
Justin Greenwood **Interest Representation in the European Union (3rd edn)**
Fiona Hayes-Renshaw and Helen Wallace **The Council of Ministers (2nd edn)**
Simon Hix and Christopher Lord **Political Parties in the European Union**
David Judge and David Earnshaw **The European Parliament (2nd edn)**
Neill Nugent and Mark Rhinard **The European Commission (2nd edn)**
Anne Stevens with Handley Stevens **Brussels Bureaucrats? The Administration of the European Union**
Wolfgang Wessels **The European Council**

Forthcoming

Ariadna Ripoll Servent **The European Parliament**
Sabine Saurugger and Fabien Terpan **The European Court of Justice and the Politics of Law**

The main areas of policy

Published

Karen Anderson **Social Policy in the European Union**
Michael Baun and Dan Marek **Cohesion Policy in the European Union**
Michele Chang **Monetary Integration in the European Union**
Michelle Cini and Lee McGowan **Competition Policy in the European Union (2nd edn)**
Wyn Grant **The Common Agricultural Policy**
Sander Happaerts and Tom Delreux **Environmental Policy and Politics in the European Union**
Martin Holland and Mathew Doidge **Development Policy of the European Union**
Jolyon Howorth **Security and Defence Policy in the European Union (2nd edn)**
Johanna Kantola **Gender and the European Union**
Stephan Keukeleire and Tom Delreux **The Foreign Policy of the European Union (2nd edn)**
Brigid Laffan **The Finances of the European Union**
Malcolm Levitt and Christopher Lord **The Political Economy of Monetary Union**
Janne Haaland Matláry **Energy Policy in the European Union**
John McCormick **Environmental Policy in the European Union**
John Peterson and Margaret Sharp **Technology Policy in the European Union**
Samuel R. Schubert, Johannes Pollak and Maren Kreutler **Energy Policy of the European Union**
Handley Stevens **Transport Policy in the European Union**

Forthcoming

Sieglinde Gstöhl and Dirk de Bievre **The Trade Policy of the European Union**
Christian Kaunert and Sarah Leonard **Justice and Home Affairs in the European Union**
Paul Stephenson, Esther Versluis and Mendeltje van Keulen **Implementing and Evaluating Policy in the European Union**

Also planned

Political Union

The member states and the Union

Published

Carlos Closa and Paul Heywood **Spain and the European Union**
Andrew Geddes **Britain and the European Union**
Alain Guyomarch, Howard Machin and Ella Ritchie **France in the European Union**
Brigid Laffan and Jane O'Mahoney **Ireland and the European Union**

Forthcoming

Simon Bulmer and William E. Paterson **Germany and the European Union**
Brigid Laffan **The European Union and its Member States**

Issues

Published

Senem Aydın-Düzgit and Nathalie Tocci **Turkey and the European Union**
Derek Beach **The Dynamics of European Integration: Why and When EU Institutions Matter**
Christina Boswell and Andrew Geddes **Migration and Mobility in the European Union**
Thomas Christiansen and Christine Reh **Constitutionalizing the European Union**
Robert Ladrech **Europeanization and National Politics**
Cécile Leconte **Understanding Euroscepticism**
Steven McGuire and Michael Smith **The European Union and the United States**
Wyn Rees **The US–EU Security Relationship: The Tensions between a European and a Global Agenda**

Forthcoming

Graham Avery **Enlarging the European Union**
Thomas Christiansen, Emil Kirchner and Uwe Wissenbach **The European Union and China**
Tuomas Forsberg and Hiski Haukkala **The European Union and Russia**

Energy Policy of the European Union

Samuel R. Schubert

Johannes Pollak

and Maren Kreutler

macmillan education palgrave

First published 2016 by
PALGRAVE

Palgrave in the UK is an imprint of Macmillan Publishers Limited, registered in England, company number 785998, of 4 Crinan Street, London, N1 9XW.

Palgrave Macmillan in the US is a division of St Martin's Press LLC, 175 Fifth Avenue, New York, NY 10010.

Palgrave is a global imprint of the above companies and is represented throughout the world.

Palgrave® and Macmillan® are registered trademarks in the United States, the United Kingdom, Europe and other countries.

ISBN 978–1–137–38883–4 hardback
ISBN 978–1–137–38882–7 paperback

This book is printed on paper suitable for recycling and made from fully managed and sustained forest sources. Logging, pulping and manufacturing processes are expected to conform to the environmental regulations of the country of origin.

A catalogue record for this book is available from the British Library.

A catalog record for this book is available from the Library of Congress.

Printed in China

Contents

List of Figures and Tables

Figures

Tables

Abbreviations

AC	Alternating Current
ACER	Agency for the Cooperation of Energy Regulators
ALDE	Associated Committee for Environment, Public Health, and Food Safety
AMBO	Albanian–Macedonian–Bulgarian Oil Corporation
AQG	Atomic Questions Group
BAP	Burgas–Alexandroupolis Crude Oil Pipeline
bcm	billion cubic metre
BP	British Petroleum
BPS	Baltic Pipeline System
BTC	Baku–Tbilisi–Ceyhan Crude Oil Pipeline
BTE	Baku–Tbilisi–Erzurum Pipeline
BtL	Biomass to Liquid
Btu	British thermal unit
CAC	Central Asia-Center Pipeline
CCME	Comité Consultatif Etats Membres Electricité
CCS	Carbon Capture and Storage
CDM	Clean Development Mechanism
CEF	Connecting Europe Facility
CEER	Council of European Energy Regulators
CHP	Combined Heat and Power Plant
CJEU	Court of Justice of the European Union
CPC	Caspian Pipeline Consortium
CSP	Concentrated Solar Power
DC	Direct Current
DG	Directorate-General
EANGP	East Anatolian Natural Gas Pipeline
EaP	Eastern Partnership
EC	European Community
ECRB	European Community Regulatory Board
ECSC	European Coal and Steel Community
EEA	European Economic Area
EEC	European Economic Community
EEX	European Energy Exchange

EMU	Economic and Monetary Union
ENP	European Neighbourhood Policy
ENTSO-E	European Network of Transmission System Operators for Electricity
ENTSO-G	European Network of Transmission System Operators for Gas
EP	European Parliament
ERD	Energy Research and Development
ERGEG	European Regulators' Group for Electricity and Gas
ESA	Euratom Supply Agency
ET	Emissions Trading
EU	European Union
EUCURIE	European Community Urgent Radiological Information Exchange
EU-ETS	EU Emissions Trading Scheme
Euratom	European Atomic Energy Community
EWEA	European Wind Energy Association
FDI	Foreign Direct Investment
GHG	Greenhouse Gas
GJ	Gigajoule
IAEA	International Atomic Energy Agency
IATA	International Air Transport Association
IEA	International Energy Agency
IPCC	Intergovernmental Panel on Climate Change
ISO	Independent System Operator
ITIP	Iraq–Turkey Pipeline
ITO	Independent Transmission Operator
JIM	Joint Implementation Mechanism
kWh	Kilowatt-hour
LNG	Liquefied Natural Gas
MEP	Member of European Parliament
MJ	Megajoule
Mt	Metric ton
MWh	Megawatt-hour
NAP	National Action Plan
NEEAP	National Energy Efficiency Action Plan
NRA	National Regulatory Authority
OB	Odessa–Brody Pipeline
OECD	Organisation for Economic Cooperation and Development
OJ	Official Journal

OPEC	Organization of the Petroleum Exporting Countries
PCA	Partnership and Co-operation Agreement
PCCE	Professional Consultative Committee on Electricity
PCI	Project of Common Interest
PEOP	Pan-European Oil Pipeline
PHLG	Permanent High Level Group
pkm	Passenger kilometre
PPC	Permanent Partnership Council
SEA	Single European Act
SGIEC	Strategic Group for International Energy Cooperation
TANAP	Trans-Anatolian Gas Pipeline
TAP	Trans-Adriatic Pipeline
TEEC	Treaty Establishing the European Economic Community
TEU	Treaty on the European Union
TFEU	Treaty on the Functioning of the European Union
TGI	Turkey–Greece Interconnector
tkm	tonne-kilometre
TPES	Total Primary Energy Supply
TSO	Transmission System Operator
TTE	Transport, Telecommunications and Energy Council
TTIP	Transatlantic Trade and Investment Partnership
TWh	Terawatt-hour
UfM	Union for the Mediterranean
UGSS	Unified Gas Supply System of Russia
UNCED	United Nations Conference on Environment and Development
UNEP	United Nations Environment Programme
UNFCCC	United Nations Framework Convention on Climate Change
UNSCEAR	United Nations Scientific Committee on the Effects of Atomic Radiation
WMO	World Meteorological Organization
WPNS	Working Party on Nuclear Safety
WWF	World Wide Fund for Nature

Introduction

Without doubt, energy security is one of the major international political issues of the twenty-first century. A glance at the annual forecasts of the International Energy Agency (IEA) from recent years (e.g. 2012: 52) reveals a projected rise in global energy demand, as much as 35% between 2010 and 2035, of which oil, coal, and gas are expected to account for roughly 60%. This is disquieting because these sources are exhaustible, geographically concentrated, and increasingly more volatile, raising the stakes of energy security as countries compete for access to resources on the one hand and search for alternatives on the other. The game is not new. Energy has been central to geopolitics at least since Sir Winston Churchill ordered the conversion of the British Navy from coal to oil. While substantial advances in recent years in renewable technologies and efficiency raised the hope that energy use could be both environmentally sustainable and politically less sensitive in the near future, other developments, including Russia's intervention in the Crimea, the unravelling of the Middle East and North Africa, and the shale revolution in the US, demonstrate that the struggle for comprehensive energy security will continue to force countries and their leaders to see access to energy resources as vital to their national security.

Nowhere is this more evident than in Europe, where geopolitics has forced the European Union (EU) to move energy policy to the top of its agenda. By the summer of 2015, after decades of keeping energy on the back-burner – a perennial issue of internal market integration – the new European Commission (2014–2019), under the leadership of Jean-Claude Juncker, was moving forward with plans to build a European Energy Union, so bundling the negotiating power of its member states vis-à-vis third-party countries, rigorously implementing the latest liberalization package, drawing up joint investment plans and introducing transparency in a notoriously opaque industry. Given the vastly different national preferences of the EU's member states, particularly as relates to liberty over national energy mixes and opposition to centralized regulation, building such an Energy Union is no small task and will only increase interest and debate on finding alternatives. Some of those traditional alternatives, such as nuclear energy or hydropower, are highly controversial because of their environmental side effects or risks. Meanwhile, newer technologies, such as nuclear fusion and carbon capture and storage, designed to clean up the toxic fumes of coal-burning, are far from market-ready. Indeed, the road from major scientific breakthrough to market viability is long (Lesage et al. 2010: 2), and anything but guaranteed.

A lot of hope is pinned on so-called renewables – a collection of allegedly clean sources of endless energy – such as wind, water, sunlight, waves, geothermal heat, and biomass. In 2010, the share of renewables was 13% of the world's total primary energy demand. While their share is supposed to grow, they are not without problems of their own. What do we do when the wind does not blow or inclement weather blocks out or diminishes sunlight? This so-called problem of intermittency needs to be addressed by traditional backup systems, running into old, well-known problems. Gas, coal, and nuclear power stations cannot merely be switched on and off at will without negative effects on the delivery of electricity and heat, essential to the functioning of modern economies and the livelihoods of its citizens. Wind farms, identified by some as a panacea for our future energy needs, tend to be small-scale and at locations quite remote from demand (Helm 2012). Large-scale installations, such as offshore wind farms and massive solar farms, require long-distance transmission grids capable of coping with naturally occurring swings in power flows, incur high maintenance costs, and require investments that consumers seem unwilling to pay. Worse, producers are equally unwilling to invest in them without assurances of receiving substantial subsidies. Add to this dynamic the questionable aesthetics of large wind farms – and in order to have a significant impact on our energy supplies those farms need to be very big indeed – and their future looks less rosy. Intriguingly, solutions, both new and old, suffer from the age-old problem of 'not in my backyard' (NIMBY).

Of course, wind is not the only renewable energy form and, in conjunction with solar and biomass, it can contribute to the 'greening' of our energy production. 'Greening' refers to the reduced CO_2, sulphur dioxide, and other climate-relevant emissions that result from energy production, distribution, and consumption.

Greening is not the same as green. One of the most important lessons of energy politics is that all energy production is dirty – be it coal, oil, nuclear, or wind-based electricity production. We all know that coal and oil are extremely detrimental to the environment, no matter whether we talk about mining, drilling, or burning, but wind turbines require fibreglass rotors, the production of which is not without health dangers, and both wind and solar technologies require massive installations to make an impact, covering spaces unthinkable in Europe's densely populated landscape. The fact is that for the time being there simply is no alternative to our overwhelming dependence on fossil fuels. The cardinal question for energy policy is then: What is the right mix of resources in order to provide consumers with a sustainable, competitive, and reliable energy supply?

As if this would not be difficult enough already, the disastrous consequences of the world's current energy-consumption patterns are no longer a scenario on a distant horizon, but a reality of the moment. Climate change, i.e. the change in temperature as forecasted by the Intergovernmental Panel on Climate Change (IPCC) and discussed in endless rounds of political talks at the United Nations Framework Convention on Climate Change (UNFCCC), is no longer a theoretical possibility, but undeniably supported by empirical evidence. Energy-related pollution directly affects the health of millions of people: outdoor air pollution kills more than three million people across the world every year, and causes health problems from asthma to heart disease for many more (OECD 2014: 11). In China alone, 750,000 people prematurely die each year from air pollution, and a recent study by the World Health Organization (WHO) estimates the economic costs of the approximate 600,000 premature deaths and of the diseases caused by air pollution in the European region at a staggering 1.6 trillion USD (WHO 2015).

Alas, most policy discussions on energy revolve around questions of the extraction, production, and trade of fossil fuels from parochial, national points of view. For example, in order to quench the rising energy thirst of its population, the Indian government was planning to increase the coal it burns in power stations, essentially adding one new coal-fired power station per week by 2016. China, which hosts a quarter of the world's more than 2,300 coal-fired power stations, foresees adding another 70 GW of power per annum, or around two large coal-fired plants per week. Research by the World Resources Institute shows that 1,199 new coal-fired plants, with a combined capacity of 1,401,268 MW, were being proposed globally in late 2013. Given the abundance of proven reserves for coal – the IEA estimates that, based on 2013 production rates, coal will last us another 3,050 years; and it will stay cheap. In fact, coal was so cheap in late 2014 that environmentally highly problematic coal-fired plants began to displace gas-fired ones, even in environmentally conscious Europe. The cause of this shift was the shale gas boom in the US, which led to the export of cheap American coal that was then willingly burned in European power stations, subsequently undermining the EU's claims that it is decarbonizing its economy.

Natural gas, which produces a little over half of the greenhouse gases emitted by coal-burning, may provide some respite in the coming decades. The international marketplace for natural gas is growing, driven by environmental concerns on the one hand and technological and market dynamics on the other. The advent of a seaborne liquid gas market, shipping Liquefied Natural Gas (LNG) between the Atlantic and Pacific basins, raises the hope that gas, like oil before it, could become a globally traded commodity, creating a myriad of opportunities for consumers such as the EU to diversify its roster of suppliers (see Chapter 7). In late 2014, the International Energy Agency predicted that LNG supplies would double in years to come, increasing the ease by which gas could replace coal. LNG is not a panacea. Europe still relies primarily on pipelines to deliver its gas at stable and affordable prices, and the current reality very much revolves around existing transport, mainly from Russia to Europe. This means that Europe's energy security is ultimately tied to the well-being of its suppliers, particularly Russia, a geoeconomic tripwire that Europe can hardly ignore. If it is to gain access to the vast resources available from the south-east – in Iraq, Iran, the Black Sea, and the Caspian – it will have to make huge investments in new pipelines. Such actions are just as likely to trigger new geopolitical competition on top of old rivalries, something evidently demonstrated by the failed planning behind the Nabucco and South Stream pipelines. Whatever the outcome of current and future plans, there is no doubt that the world's energy flows are changing, and the question of whether or not natural gas will become the 'bridge fuel' for Europe's plans to build a low-carbon future remains to be seen.

EU efforts towards building the low-carbon economy of the future have been underway for a while now. Indeed, one of the more remarkable features of the evolution of EU energy policy has been the changing entrepreneurial role played by the European Commission, which has maximized the tools at its disposal to overcome divergent national preferences en route to forming a common European position. While geopolitical forces now enable the Commission to take advantage of treaty obligations for energy solidarity and to prepare a more active role for the EU on the external front, it first relied on competition rules to liberalize its internal energy markets before turning to environmental regulation to tackle the multidimensional issues of sustainability and climate change. It has done this by setting ambitious energy and climate objectives, the latest one being the so-called 2030 objectives (COM(2014) 15 final) that include a 40% cut in greenhouse gas emissions (compared to 1990 levels), a 27% share of renewable energy consumption, and an increase in energy efficiency. Those internal objectives underpin the EU's international or 'external' leadership claim in protecting the world's climate. An ardent supporter of international agreements, from the Kyoto Protocol (1997) to the Bali Road Map (2007), the Cancun Agreements (2010), the Durban Outcomes (2011), the Doha Climate Gateway (2012), and finally the Warsaw Outcomes (2013), Commission officials have racked up countless air miles in pursuit of curbing the world's energy hunger, and especially its greenhouse gas emissions. While the Commission deserves credit for good effort, the results so far give cause for concern. The IEA's *World Energy Outlook 2013* estimated a growth in energy-related CO_2 emissions of 20% by 2035, leaving the world on a trajectory consistent with a long-term average temperature increase of 3.6°C, far above the internationally agreed 2°C target (IEA 2013: 24).

As bleak as these global circumstances may seem, this book does not focus on why so many energy policies seem to lead nowhere. Rather, it concentrates on the development, the actors, processes, and main challenges that the European Union faces under the aforementioned circumstances. Dedicated to the establishment of a common market (Article 3 TEU), the protracted development of the internal energy market was and remains remarkably slow. By looking into the historical background and by charting national preferences, this book attempts to provide an explanation for this slow development. It provides the reader with a clear picture of who does what in the formulation of the EU's limited energy policy, and it analyses the legal and political framework of this important policy field. It charts and explains the difficulties of bringing together 28 veto players with widely divergent interests, players that sometimes defy economic or political logic, and certainly defy the spirit of the treaties to which they are committed. It shows a European Commission devoted to ambitious goals, but largely powerless to act in the face of member states led by political elites whose objectives are dominated by short-term considerations rather than by long-term strategic planning. Yet it also shows how that very same Commission has used cunning reason and entrepreneurship to the benefit of the member states as a whole and, thus, led the way in driving towards a common EU energy policy.

One of the main problems of analysing energy policies in general is the complexity of the subject matter. Although we use energy on a daily basis, we tend to forget the basic physical determinants we learned in school. Remembering basic physics, however, makes it easier to follow discussions about substitutes, renewables, workload, etc. We therefore provide in the following paragraphs, a quick refresher-type course, taking key energy definitions from the Glossary of the International Energy Agency (2015) and the Energy Information Agency (EIA 2005) of the US Department of Energy.

Simply stated, energy is the capacity for doing work such as lifting, accelerating, or heating, and measured by the capability of doing work or the conversion of that capability into motion. In a world void of energy, there is simply no activity. In order to grasp the choices and challenges faced by consumers and policymakers alike, it is necessary, therefore, to remember five important facts.

First, energy comes in different forms, which we can place in two overarching categories: kinetic and potential energy. Kinetic energy refers to motion and can be found in things that move, such as wind and flowing water. Included in this category are also radiant, electric, and thermal energy. Potential energy refers to stored energy and includes chemical, stored mechanical, nuclear, and gravitational energy. Chemical energy is stored in the elements of which all things are comprised, and are usually released through the process of burning. Petroleum, natural gas, coal, wood, and biomass (including liquid biofuels) all contain significant amounts of chemical energy. Nuclear energy is the energy, stored within the nucleus of an atom, that holds the atom together, and gravitational energy is the energy of position. The most common form of gravitational energy use is the water reservoir behind a dam used for hydropower.

Second, energy can be converted from one type to another, also known as the law of conservation of energy. Fire, for example, is the result of chemical energy converting into thermal and then radiant energy. The chemical energy found in wood and oil is the result of photosynthesis, which itself is a process plants use to convert radiant energy (sunlight) into chemical energy. Moving water or steam is used to drive turbines, creating mechanical energy, and wood or gasoline is burned to convert water into steam and pressure. Energy is thus never lost, only transformed from one form to another. The rate at which we extract and use energy is called energy efficiency.

Third, not all natural resources store energy equally. Gross (higher) and net (lower) calorific values of a resource depend on its composition. Using rough comparisons of energy density, for example, one kilogramme of petroleum has an energy density of between 42 and 45 Megajoules (MJ), which is comparable to circa 29 MJ for cooking coal, 17–18 MJ for herbaceous biomass, and circa 20 MJ for farmed tree wood. Purely from the perspective of efficiency, it is more energy-efficient to burn a kilogramme of oil than wood. This distinction is extremely important when evaluating the economic viability of energy resources. Measurements for comparison are made using calorific equivalents, among which are calories, joules, and British thermal units. Other measures often cited, such as watt-hours, refer to the output power and can vary greatly depending on the process used.

Fourth, natural resources containing energy come in a wide variety. Some come in the form of liquids, such as oil and water. Others come as solids, such as coal and wood, or as gases, such as wind (the movement of air, which is comprised of gases) and natural gas, while only one, sunlight, comes in the form of pure energy. Since all energy is convertible, electricity can be made from all of them, and thus electricity is sometimes referred to as a converted energy.

Fifth, some energy-containing resources are finite, while others are not. Finite energy resources are called depletable resources, and refer to those materials that require millions of years to create, such as coal, petroleum, and natural gas. Renewable resources can be replenished in a short enough period of time for economic decision-making, and include biomass, geothermal, wind, and solar.

As illustrated above, energy can be found in different resources, stored in different forms, and converted. Throughout history, human understanding of how to extract and use energy has evolved, and with that evolution so too has its economic organization. Only a hundred years ago, most energy production and consumption in Europe were local events. Coal was mined and wood chopped then shipped the shortest possible distance by horse, mule, or river, to be used for heating, lighting, and basic machinery. In the early nineteenth century, Europe relied heavily on coal for heat, on horses for domestic transport, and wind and flowing water for milling. Yet a century before that, international trade largely depended on wind and mule power; and a millennium ago the single largest provider of energy was none other than forced human labour.

Throughout history, energy has been a core aspect of social and economic change. In less than two hundred years, human societies have moved from energy independence to energy interdependence. Energy use shifted from essential consumption to mass consumption, and from renewable resources to depletable ones. Today, the energy we consume in our homes and the fuels we use to shuttle between locations largely come from somewhere else. It is precisely the challenge of acquiring, delivering, and converting those resources most efficiently that is the driving force underlying the formulation of an energy policy. It is a particularly cogent one for the European Union. This book specifically addresses how the European Union and its member states go about (and have gone about) acquiring and distributing the energy its citizens need in a manner that is concurrently economically fair, affordable, and environmentally sustainable. It is a story about a long, arduous, and unfinished road from ferociously independent, national security-driven post-war reconstruction to an increasingly integrated market and hopes for solidarity in foreign energy relations.

The book proceeds as follows. Chapter 1 introduces the EU's energy portfolio, a complex and diverse set of economic, technical, and political interests that shape the EU's energy policy. Three dimensions – internal, external, and multidimensional – are used to organize the EU's energy policies within the broader context of the pursuit for comprehensive energy security. These two organizing principles (policy dimensionality and the pursuit of comprehensive energy security) are referenced throughout the book. The building of the internal market, energy security, and the fight against global warming, what the EU respectively labels competition, security of supply, and sustainability, are the corresponding policy areas that are analysed in detail.

Chapter 2 lays out the basic principles and specific terminology of energy economics. One easily forgets that behind the seemingly innocuous switching on of a light, there are multiple companies and a vast network of huge machinery involved, working incessantly, to deliver that service. It involves a complex chain of activities that runs from exploration to mining, from refining to distribution, and from utility providers to households and businesses. A special focus is placed on the three end uses of energy – electricity, transport, and heating – each of which has specific characteristics and relies on different mechanisms and combinations of energy production.

Chapter 3 reviews the protracted struggle for the integration of nationally organized energy markets. Energy issues were central to the very foundation of the European integration project. Two of the three founding treaties, the European Steel and Coal Community Treaty (1951) and the Euratom Treaty (1957), were explicitly related to energy. However, it was only with the passing of the Lisbon Treaty (2007) that energy policy found its own chapter in primary law. This history can be understood as a tug of war between the Commission and the member states, whereby, in the face of the latters' resistance, the former turned to environmental regulation as a driver to advance a common energy market. That it took decades to arrive at only a partially integrated internal energy market demands an explanation. Why are the member states so reluctant, if not outright opposed, to adopt a common energy policy? The chapter also charts the different theoretical explanations that address this question by looking at the shifting distribution of competences, the changing nature of the legislative output, and the contribution of the major theories of integration. This leads directly to the question: Who does what in European energy politics, which is the subject of Chapter 4.

How the attempts to forge a joint regulatory framework for the three end uses (electricity, heating, and transportation) panned out between the EU member states is the subject of Chapter 5. The instruments of the EU are mostly regulatory. It is a basic feature of the integration process that the market will drive producers and consumers towards Pareto-optimal outcomes. Unfortunately, market distortions are the rule in the energy domain, and not the exception. The character of energy as a strategic good, the member states' different levels of dependence on foreign energy resources and suppliers, and their diverse geographical conditions each impact on the protracted struggle to forge an internal energy market.

Since Europe's energy market is overwhelmingly dependent on oil, gas, and coal imports, international crises have huge repercussions for EU policymakers and consumers alike. Thus, during times of crisis (e.g. the oil crises in 1973 and 1979, the Gulf War in 1990, and the annexation of the Crimea by the Russian Federation in 2014), Europe's political leaders unite under the rally cry of reducing import dependence. Despite increasing storage facilities, cross-country interconnections, reverse flow capabilities, etc., one can hardly mask Europe's external dependence. One of the strategies to reduce that dependence is to invest in renewables and increase energy efficiency, questions dealt with in Chapter 6, which also charts the EU's attempts to foster these and other new forms of energy production.

Chapter 7 sheds light on the highly sensitive nature of international energy politics. Relying on a mix of diplomacy, rule export, and economic statecraft, the EU tries to manage and mitigate the risks associated with energy interdependence and market volatility. So far, those efforts have been limited mostly to political rhetoric, and often in the aftermath of international crises. The EU's plan to build an Energy Union holds the promise of strengthening solidarity, bundling the external negotiating power of its now 28 member states, and further integrating key aspects of its internal energy market. However, forging a common external energy policy to deal with the energy challenges of the twenty-first century is a target the EU has so far clearly missed.

The book concludes by addressing the main internal and external policy challenges faced by the European Union. Looking at selected policy cases, we demonstrate the links between the EU's internal, external, and environmental policies. Looking internally, we analyse the challenges of furthering the internal energy market, assess the future of European energy solidarity and the Energy Union, and examine the practicality of the goal to increase the role of renewables in its energy mix. We also point out the urgent need for increased investment in Europe's energy transmission infrastructure, investigate the future role of indigenous supplies, and distinguish between the 'hype' and reality of hydraulic fracturing (fracking) in the EU. Turning to external geopolitical challenges, we discuss the connections between the EU's future relationship with Russia, the development of a Southern Corridor, and its complex relationship with Turkey. We also appraise the value of accessing Caspian resources and gauge the sensitive need to responsibly exploit the Arctic. Positioning the EU's current and impending energy challenges in this way serves an important point. By coalescing the internal, external, and multidimensional aspects of EU energy policy and showing how they are inextricably linked, we demonstrate just how important the promise of a single internal energy market is to achieving the EU's overarching goal of achieving comprehensive energy security for its citizens, a goal that is concurrently closer than ever and yet still so far away.

Chapter 1

The EU's Energy Portfolio

Energy policies encompass crucial public policy decisions over the production, distribution, and consumption of energy, from incentives, taxes, and regulations to the establishment of strategic national security goals. Every decision affects specific interest groups differently and can significantly affect a country's economy. Politicians, lobby and interest groups, environmental activists, and also industrialists argue over which resource to use, as well as whether, when, and how to switch to alternatives. Governments order their citizens to change their energy-use habits, buy new light bulbs, and drive cars that are more fuel-efficient. Determining the optimal policies to deliver the best possible political and economic outcome is no easy task. Should governments just leave energy prices to the markets or tax energy use and invest the receipts in alternative resources and technologies in the hopes of strengthening domestic markets and reducing dependency on external supplies? Even if one uncovers the best approach for a particular state, can one also conclude that the approach is equally valid for a group of nations as diverse as the European Union?

Despite explicit declarations and titles, energy policies often tend to be rather vague and diffuse. Relevant decisions largely have an internal focus, just like most other policy areas, but unlike most other policy areas, internal energy choices carry serious external implications because they directly affect national and economic security. It is, therefore, no surprise that the EU's energy portfolio consists of a mixed bag of complex, interrelated, and sometimes contradictory policy dilemmas.

In its broadest sense, EU energy policy can be described as the pursuit of comprehensive energy security. To achieve this, the EU pursues three broad objectives: mitigating climate change (known commonly as sustainability), guaranteeing that energy is affordable for European customers, and achieving security of supply. Its methods to achieve these three objectives are, respectively: reducing carbon dioxide emissions and increasing energy efficiency; enhancing competitiveness through a functioning internal energy market; and building an infrastructure of intra-European interconnections, while diversifying both internal and external suppliers and reducing the negative consequences of import dependency on third parties, through solidarity and cooperation.

As the EU is the third largest consumer of energy worldwide, behind China and the US, (13.8% of global inland consumption), its economic prosperity depends on market forces that are largely beyond its control. However, those market forces are not beyond its influence. The Union took many measures over previous decades to reduce demand for particular fuels and increase demand for others. It achieved this through legislative and regulatory measures that prioritized efficiency, environmental concerns, and cooperation. According to the European Commission, for example, EU energy efficiency standards introduced as part of its 20-20-20 targets in 2008 (European Commission 2008) played a role in reducing overall EU energy consumption by 3.8% between 2005 and 2010 (European Commission 2013). Yet the actual role of EU targets in this process remains open to debate: economic cycles, the downsizing of global economy, the closure of energy-intensive industries – especially in Eastern and Central Europe – and mild winters also heavily affect energy consumption. Furthermore, the EU's energy policy may be having some significantly negative strategic consequences. As Dieter Helm (2012a) notes, 'Europe has been deindustrializing too, and this has also encouraged energy-intensive production to move overseas'. In other words, Europe may be consuming less energy and producing fewer emissions, but in the process is simply pushing that production and pollution to other parts of the world. In fact, it appears that the US is moving past Europe in environmental terms, largely because it has eschewed a top-down energy policy and left the market to determine the optimal path to sustainability instead.

The EU's energy agenda is not static: it expands as new issues arise, and some of the issues are quite contentious. For example, the EU is split over long-standing questions about national vs. community-wide energy mixes, its huge dependence on foreign suppliers for natural gas – Russia in particular – and the continued use of nuclear power. These debates manifest themselves in national policy choices that often lead to contradicting results. For instance, despite its policy to become greener and less dependent on external suppliers, Germany decided to phase out its nuclear power facilities only to experience a sharp rise in both its coal consumption as well as its dependence on Russian gas. Likewise, despite the clear environmental focus of the EU, supporters and critics of fracking expose the perennial dilemma between the aim to reduce external dependency and the need to secure the environment.

Less contentious issues are no easier to resolve. While the EU has successfully exported energy norms along its periphery, it has been less successful maintaining unity on energy relations with Russia or on developing alternative southern routes around Russia, the flagship project being the abandoned Nabucco pipeline project. Because energy security is no less a national concern than it is a Union matter, almost every attempt by the Commission to secure jurisdiction over any aspect of external energy policy runs into conflict with member-state interests. Increasing external policy coordination is a tough sell, whether it is presented as developing and sustaining a coherent, multilateral negotiating position or enacting supranational oversight. For example, discussions about a possible consultation role for the Commission in bilateral energy-supply agreements (European Commission 2011g, 2011j) – something that would exemplify European solidarity and coordination – stirred up traditional member-state concerns about sovereignty (de Jong 2012). In short, energy is a touchy subject, and the EU's energy portfolio is crowded with important, complex, and controversial issues that sometimes scratch at the political heart of the European Union. Indeed, regardless from which angle one looks at energy politics, political decisions are extremely complex, involve a high number of variables, and are based on very long timelines.

This chapter introduces readers to the key energy policies and debates in the European Union, what we call the EU's energy portfolio, and lays out a typology to organize and analyse them. We begin by introducing a typology of policy areas, a framework in which we distinguish between internal, external, and multidimensional domains. After a brief handling of each of the aforementioned domains, we derive key policy areas or policy groups within which we can position the most important energy-related policies. These policy groups form the EU's energy portfolio. We follow with an in-depth look at the collective and overarching goal that both exemplifies and defines the EU's energy portfolio, namely the notion of comprehensive energy security.

This chapter also serves as a launching point for more detailed explanations found in later chapters. Within the EU's energy portfolio, one can find such diverse matters as the use of nuclear power (specifically discussed in this chapter), the building of a common internal energy market (Chapter 5), the coordination of overall greenhouse gas emissions (Chapter 6), and the exporting of EU rules and norms about energy contract transparency (Chapter 7).

A typology of the EU's energy policy

Analysing EU energy policies requires that we first understand the context of the stages on which they play; the EU is after all a collection of independent sovereign states operating under a series of treaties that transfer jurisdiction in certain areas to a supranational centre. No aspect of an energy policy can be considered entirely internal or external; in most cases, choices in one area have a significant effect on the other. Still, issues such as regulatory politics or drilling and mining rights are clearly more internal than external, just as coordinating supply contracts with third parties is more external than internal. Hence, we need to distinguish between those policies that are primarily internal, those that are primarily external, and those that span both categories. Table 1.1 presents a broad typology of the EU's energy policies as described above. The typology serves to organize and simplify the complex set of issues that constitute the EU's energy portfolio.

Using the EU energy-policy typology is a good way of thinking about these issues. It provides a framework for allowing students and researchers to understand how each element of EU policy relates to others, and helps us to identify and select critically important policy areas, the choices within which involve substantial debates. Internally, there are three areas of focus: the building of a common internal energy market, debates over the status of and support for nuclear power, and the search for substitute fuels through research and development. Externally, the EU focuses on developing and coordinating a common external energy policy through rule export and energy diplomacy. It struggles with contentious pipeline politics, debates over the construction and development of LNG ports and facilities, and the enormous task of reducing its extreme external energy dependence. Finally, the internal and external policy dimensions coalesce in the area of climate change, an area that has served as both a tool and a goal to achieve its

Table 1.1 *EU energy policy typology*

Internal	*External*
Establishing the internal energy market - Common market principles - Liberalization - Deregulation and re-regulation - Intra-EU networks and infrastructures - Harmonizing energy taxes - Subsidies *Nuclear power politics* *Developing new energy technologies* - Research funding - Subsidies, feed-ins, etc.	*Common external energy policy* - Reducing external dependency - Ensuring secure energy supplies from abroad (energy supply security) - Diversifying suppliers/ supplies - Pipeline politics and LNG ports - Rule export/energy diplomacy
Multidimensional	
Reducing greenhouse gas emissions (climate change)	

internal and external energy policy goals. Each of these policy areas is addressed in subsequent chapters of this book.

The most well-known and tested policies fall within the internal domain. Every European consumer is familiar with the debates around the use of nuclear power, fuel efficiency standards for automobiles, power efficiency standards for light bulbs, and the price effects of increased renewable electricity production. The success of these policies depends in part, however, on the success of other key policy goals that largely lie outside the control of the EU and its member states. These include: ensuring secure energy supplies from abroad (energy supply security); diversifying the roster of suppliers (supplier diversity); reducing greenhouse gas emissions (climate change); developing new energy technologies through research funding (substitution); continuing the discussion on the costs, merits, and risks of nuclear power; and coordinating a common external energy policy.

Likewise, reducing dependence on external energy suppliers is not a matter of external energy policy alone. Changes in efficiency standards and the increased use of renewables, as well as the use of nuclear power sources and advances in technology, also play important roles in reducing demand for key imported fossil fuels, oil and gas. Structural changes in the international energy market can change the dynamics of internal demand and supply. The maturation of the shale oil and fracking industries in the United States and a boom in natural gas transportation (LNG) could 'push oil to a structurally lower price' (Riley et al. 2012: 29) and, thus, sap much of the economic demand for alternative fuels similarly to how cheap and abundant coal ate into natural gas' share of electricity production in Germany in 2013 (Burger 2013).

The phenomenon of global warming creates added pressure to find alternatives to fossil fuels, particularly in the EU where greenhouse gas emissions 'increased by 33% during the period 1990 to 2007' (European Commission 2014: 14). Besides inciting Europe to seek reductions in its CO_2 emissions by almost 20% of its 1990 levels by 2020 and 40% by 2050, climate-change concerns are leading the EU to act both directly through legislation and indirectly through member-state research budgets. The European Commission's *Energy 2020 – A strategy for competitive, sustainable and secure energy* (2010a) and its *Energy Roadmap – 2050* (2011b) highlight these priorities. Thus, Europe is working diligently on a wide range of energy-saving or alternative-energy projects, including among others wind power, carbon sequestration, safer nuclear power plants, new forms of hydroelectric energy production along riverbeds and coastal planes, and photovoltaic and solar-thermal power generation plants.

The internal dimension

The internal dimension of energy policy engages a diverse set of actors, institutions, processes, and instruments. Many of these are spatially and temporally specific to the political–economic structures and conditions of the member states. The single most important and established policy area is the creation of a fully functioning internal energy market. Already underway, the completion of such a market includes diverse policies such as promoting common-market principles, implementing programmes of liberalization, national deregulation and supranational re-regulation, developing intra-EU energy networks and infrastructures, and harmonizing energy taxes and subsidies. These are policy areas in the EU energy portfolio that one can claim are within the control of the EU or at least within the control of its member states.

The external dimension

The external dimension is less complicated in terms of the number of factors involved, but is an area where it is much more difficult to reach consensus between competing national interests, historical relationships, and respective geographic positions. The external dimension of EU energy policy can be summarized as the pursuit of a common external approach to reduce the Union's dependence on foreign suppliers. Its substantial imports from a limited number of suppliers for most of its oil and gas create vulnerabilities to external supply shocks, disruptions, price swings, and political agendas, all of which negatively affect solidarity among member states and economic prosperity within them.

The multidimensional domain

Some policy areas naturally span both the internal and external dimensions. The most important one is climate policy, which focuses on the Union-wide reduction of greenhouse gas emissions and the environmental sustainability of industry and the economy. Because energy production, transmission, and consumption create negative externalities that transcend borders, it is impossible to successfully implement environmental policies solely by internal legislation and regulations, or even externally by multilateral treaty. For environmental strategies to work, changes in energy and industrial production must be universal: international agreements and coordination are required. For these reasons, EU climate policy constitutes a special multidimensional area.

One may even consider EU climate policy as the central pillar of EU energy security, because if the member states collectively meet their ambitious greenhouse gas targets, the EU could achieve the delicate balance between economic prosperity and environmental sustainability for which it is searching. The most obvious manifestation in policy terms of the EU's climate policy is not, however, the targets its sets for reduced greenhouse gas emissions per se, but it is rather the combined effect of reduced external dependence through the increased use of renewables, increased efficiency of appliances, and greater integration of the member-states' energy infrastructures at competitive prices.

The real target: Comprehensive energy security

The ultimate measure of success for the EU energy policy is the realization of comprehensive energy security (CES), namely that affordable supplies of energy in whichever form they are required, meet current and expected demands from household consumers and industry alike, with enough built-in redundancy, alternatives, and substitutes to minimize the negative effect of disruptions irrespective of the cause. The formula to achieve this was laid out by the Commission in a 2006 Green Paper that identified sustainability, competitiveness, and security of supply as the three pillars of EU energy policy (European Commission 2006). It was later established as policy through a Commission Communication called *An Energy Policy for Europe*, in which the issue of collaborating to achieve these three pillars was described as a project that would return the EU 'to its roots' and initiate a 'new industrial revolution' (European Commission 2007: 3–5). At the heart of achieving these bold strategic goals was the establishment of a 'real internal energy market', one where security of supply is achieved as energy supplies and services move seamlessly between member states and are delivered promptly and efficiently anywhere within the Union where they are needed. One group of EU energy policy observers has described the EU's approach in terms of three 'mantras' attached to key events: the green issue tied to Kyoto, the market issue tied to the Lisbon Treaty, and the security of supply issue tied to Moscow and Europe's considerable dependence on Russian gas supplies (Glachant et al. 2010). For purposes of consistency and clarity, we will use the EU's language of pillars to elucidate the categories and details of its energy portfolio.

The underlying logic of the three-pillar policy is sound in itself. Real competition should lower energy costs for citizens and businesses and stimulate investment. Tough environmental legislation and the emissions trading system, both of which aim at limiting greenhouse gas emissions, should make economic expansion and the delivery of the necessary energy services more sustainable over time, by promoting increased use of renewable energy technologies, combined heat and power and microgeneration, and other non-conventional supplies. Finally, security of supply should protect European consumers from some of the negative consequences caused by international price volatility (in oil in particular), as well as avoiding economically crippling brownouts and blackouts in the electricity sector. Together, these three pillars, if implemented in their entirety, would create a functioning internal energy market. That market in turn would be competitive, robust, redundant, and interconnected. It could increase or reroute excess electricity during hot summers to affected areas and would contain enough strategically positioned oil and gas storage facilities, both in quantity of installations and volumes of storage, to respond to emergencies. It would also be able to offset interruptions in supplies, e.g. gas from Russia or oil from the Middle East, through an interconnected infrastructure that could move the necessary energy resources from any unaffected area in the Union to where it was needed at short notice.

The Commission's focus on competitiveness as a key aspect of comprehensive energy security is rooted in three key concerns that increase the risk of shortages and keep prices high. Specifically, these include overexposure to price volatility in international energy markets, the uneven concentration of the EU's external oil and gas providers, and internal oligopolistic control of energy generation and distribution. Increasing competitiveness, essential to a functioning internal energy market, should lead to a stronger investment climate, and this in turn should lead to job creation. The resulting combination of increased investments, jobs, efficiency in consumer appliances, and energy generation and distribution should not only reduce energy costs but also enhance both living standards and quality of life for the average European citizen.

Parallel to its approach on competitiveness, the Commission continues to profit from its long experience in using environmental legislation to drive its energy policy. Ever since adopting the *United Nations Framework Convention on Climate Change* (the Kyoto Protocol) in the late 1990s, the EU has struggled to find a balance between achieving its emission reduction goals and maintaining industrial growth and competitiveness. The EU positions itself as world leader in the drive to respond to and mitigate the effects of climate change. Yet, structural changes in the international energy market, particularly the increased availability and use of cheap coal for electricity generation in Europe, have eaten into the legitimacy of that claim, as its carbon consumption was on the rise in late 2013 (Helm 2013).

The EU's goal of achieving collective reductions of just above 5% (of 1990 levels) by 2012 led to the establishment of the EU Emissions Trading Scheme (ETS). In 2007, the Commission proposed, and in 2009 passed, its *2020 Climate and Energy Package*, known as the 20-20-20 plan (European Commission 2008a), the targets from which became binding through the *Renewable Energy Directive* of the Third Legislative Package. The plan, which set the goals to reduce greenhouse gas (GHG) emissions by 20%, increases the share of renewables by 20% and enhances its energy efficiency by 20% all by 2020, and quickly became the motto of EU energy policy. The package was intended to move Europe to a so-called low-carbon economy and promote green growth and jobs by promoting four pieces of complementary legislation. It began by reforming the EU-ETS: with the start of the third trading period in 2013, the EU introduced a single EU-wide cap on emission allowances to replace the less efficient system of national caps, and intended to cut the cap annually in order to reduce emission levels to 21% below the 2005 level by 2020. Existing plans include introducing a system of auctioning, beginning with the power generation sector, to replace the original system of free allowances. As the ETS only covers circa 40% of EU emissions, the member states also agreed, under the *Effort Sharing Decision* (European Parliament and European Council 2009) to binding reduction targets for sectors not covered by the ETS, including agriculture, waste, and the essentially important transport sector. Interestingly enough, the latter originally excluded the aviation industry, but it too joined the fold in 2014. Unfortunately, no agreement has been reached on emissions from land use and forestry. Still, the Commission claims that member-state obligation to report annual emissions under a newly instituted EU monitoring mechanism is reason for optimism. It believes that this obligation will increase the likelihood that the EU will achieve its targets. However, few of the member states are on course to meet their own national targets (Helm 2005; Helm and Hepburn 2010).

Through its *Renewable Energy Directive*, the EU set binding national targets increasing the share of renewables in their energy mix (EP/Council 2009). This step should not be underestimated given the enormous resistance that member states have shown historically against any supranational limitations on their energy mixes. Resistance was overcome by setting different targets for member states based on their starting points and potential: for example, Malta has a target of 10%, whereas Sweden's is set to 49%, both by 2020 (see Figure 6.1). The final element of the climate and energy package included establishing a legal framework for carbon capture and storage (CCS) technologies (EP/Council 2009i).

The third pillar of the EU's approach, namely security of supply, remains its weakest link. In a November 2000 Green Paper, *Towards a European Strategy for the Security of Energy Supply*, the European Commission painted an alarming picture of the future. It predicted that '[i]f no measures are taken, in the next 20 to 30 years 70% of the Union's energy requirements, as opposed to the current 50%, will be covered by imported products'. Calling the economic consequences of this dependence 'heavy', it recognized that the EU lacks 'the means to change the international market' (European Commission 2000). In 2003, the EU again noted that the need to reduce its dependence on imported energy was essential to the security of energy supply in the medium and long term (EP/Council 2003a) and reconfirmed that message in a 2007 Energy Policy Communication (European Commission 2007).

Little changed in the decade that followed the Commission's initial warnings; indeed, it worsened. Between 2002 and 2013, the EU's dependence on imports grew: from 47.5% to 53% for all fuels, from 33% to 44% for solid fuels, from 76% to 87% for oil, and from 51% to 65% for gas (Eurostat 2015). Denmark was the only country that exported more energy than it consumed over the period, but nonetheless saw its independence, or more accurately energy autonomy (i.e. its positive overall energy balance), drop from almost 42% in its favour in 2002 to less than 4% in 2012. Simply stated, despite all its efforts and plans, the EU, both as a whole and in its parts, grew more dependent in the first decade of the twenty-first century (see Figure 1.1).

Insofar as one considers external dependence a measure of energy insecurity, Europe's heavy dependence on external energy supplies constitutes a serious risk to its domestic economic and social welfare. This increasing dependence raises the stakes and risks associated with commercial disputes, as was the case involving Russian gas over disputes with Ukraine, or political unrest and uncertainty, as has been the case in Nigeria, Algeria, and the Middle East. Even if the EU were to strengthen its external energy relations protecting Europe from transit-related disputes, build up new networks of pipelines from Russia or Central Asian and Middle Eastern sources via Turkey, increase its LNG-import capacity, and link up its own

Figure 1.1 *EU energy dependence on all products, by per cent (2012)*

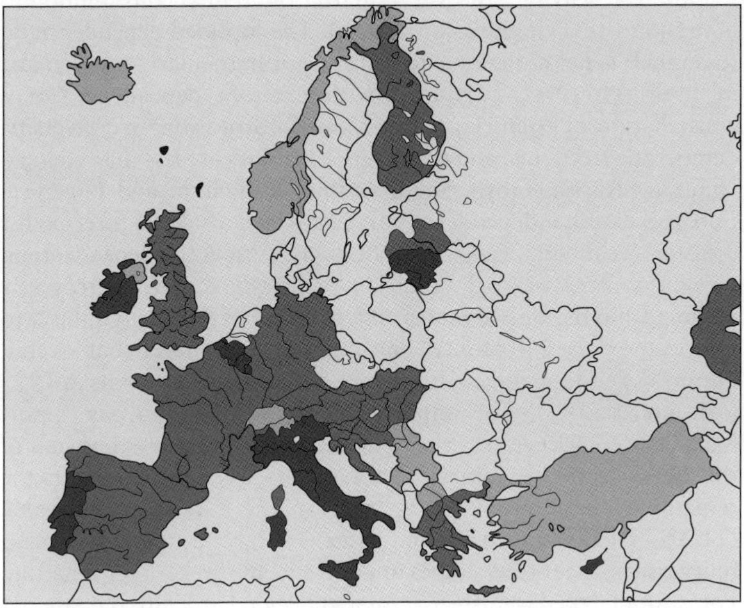

Legend

■ 11.9 – 42.6333 ■ 42.333 – 73.3667 ■ 73.3667 – 104.1

Minimum value:11.9 Maximum value:104.1

Source: Data from Eurostat (2015a)

national and regional oil and gas networks with externally sourced import lines, accidents or technical failures on a major import pipeline would still have severe consequences. The recognition of this problem explains in part why the Commission is increasingly arguing for the controversial position that there should be community-level coordination to diversify both suppliers and transit routes as well as to take an active role in contracts.

To put the EU's energy import dependence into context, it is helpful to compare which fuels are used and by which sectors in Europe. Of all the energy consumed in 2011, 40.3% came from oil, another 21.9% from natural gas, 4.4% from solid fuels, and 21.6% from electricity; just a little over 7% came from renewables, 4.4% from derived heat, and 0.5% from industrial waste. Meanwhile, the transportation (33%), industrial (26%), and service (12.7%) sectors accounted for roughly 72% of all energy consumed in the EU in 2011, leaving just over a quarter of all final consumption to households (EU Commission 2013b). The lopsided dependence on fossil fuels is neither unique to Europe nor particularly problematic in itself. However, combined with external dependence on a constellation of countries where the record of economic diversity, democratic freedoms, and even regime stability are less than stellar – including Russia, Libya, Saudi Arabia, Iraq, Iran, and Nigeria – Europe's external dependence can easily be considered a recipe for disaster (Schubert 2007). The EU was even 62%-dependent on foreign sources of coal in 2011. Although coal is increasingly coming from the United States and, thus, is not especially subject to politically caused volatility, one must bear in mind that overall import dependency on coal is three times as much as it was in 1995 (just under 20%). The importance of the coal and gas import balance becomes evident when one considers that more than half of the EU's electricity comes from natural gas and coal, with the remaining share provided by nuclear (27.6%) and renewable (21.3%) energy; and while the latter is subject a number of reliability issues, the former comes under frequent threat of termination. For example, the Swedish government decided in 1980 to phase out nuclear power, only to repeal this policy in June 2010. Likewise, the German government, under a Social Democratic and Green coalition, decided in 2000 to phase out nuclear power by 2022, then put it on hold in 2010, before recommitting to the phase-out a year later in the wake of the Fukushima disaster.

The EU's considerable dependence on external energy resources points to two of its largest challenges lying ahead: mediating a stable supply of fossil fuels, and finding technological alternatives and substitute energy sources. However, its dependence is not a one-sided affair: many of the EU's current suppliers need the EU's financial power, and in the case of oil, its concentrated refining capabilities. In fact, the OECD's 30 members provide more than half of the world's oil refining capacity; almost a third of the world's refining capacity, that is the ability to turn oil into useable fuel, is located in the United States (18%) and the EU (16%). Despite being massive producers and exporters of oil, Russia and the remaining states of the former Soviet Union lack substantial refining capacity (6.2% and 2.2.%, respectively, or less than 9% of global capacity) (BP 2013). Thus, the negotiating power inherent in the imbalance between oil producers and consumers is largely offset by the global imbalance in refining capacity and leaves both supplier and importer inextricably bound in a relationship of interdependence; and that turns out to be good news for the EU's security of supplies.

Still, energy overwhelmingly defines the trade and investment relationship between the EU and Russia. Despite a decline in the EU's dependency on external sources for gas over the last ten years, its foreign dependency poses a strategic security risk to the Union's economy should any major disruption occur either by accident or design. Within this context, Russia is the most important provider of gas to the EU, supplying around 30% of all gas bought and sold in Europe, and about a quarter of all gas consumed in its economy. However, EU members are not equally dependent on Russian gas: while Western states such as the UK (0%), Spain (0%), and France (17%) are either not at all or less dependent on Russian gas supplies, large economic players such as Germany (40%) and Poland (54%), and many of the newer and smaller Central and Eastern members are between 50 and 100%-dependent, making the EU vulnerable to divide-and-conquer tactics in its relations with its eastern neighbour. Yet the EU has an edge that is rarely talked about, and one that quite significantly changes the nature of the discussion about EU dependence on Russia when taken into context. EU purchases of Russian gas 'are vital to the ruling Russian elite' (Ratner et al. 2013). Most EU trade with Russia in terms of the flow of Euros is in the fuels sector, and FDI data reveal that the EU is investing far more in Russia than vice versa. As a result, while approximately half of the EU is dangerously dependent on the stable flow of Russian gas supplies, Russia is economically dependent on EU investments and trade, with the EU accounting for some 34% of all imports into Russia and 45% of its export revenues in 2012, giving leverage to the Union, if it can act as a united force (European Commission 2013b). Indeed, concerns that Russia's influence is increasing appear to be unwarranted, at least as far as overall supplier diversity is concerned. Russia's share of overall EU member natural-gas imports has been declining: in fact, Russian imports were 10% below their 2011 level in the first quarter of 2013 and, by comparison, '[p]hysical flows from Russia remained below the volumes registered in the first half of 2011' (European Commission 2013c: 1). Understood in the context of the EU's increased use of gas to generate electricity, its addition of LNG terminals, and effective efficiency measures, the EU has successfully implemented a process towards diversifying its natural gas suppliers, albeit more as a result of using alternative sources, including coal and deindustrialization.

The energy relationship between Russia and the EU is therefore as much a story of interdependence as dependence, and this fact is pivotal in understanding the stability and strategic nature of EU–Russian relations. The fact is, despite their vastly different approaches, both share the common interest of maintaining reliable supplies, and, overall, Russia has proven to be a reliable supplier.

Interdependence is, nevertheless, no remedy for the depth of the EU's external energy dependence. Its member states and the Union as a whole are actively engaging in projects to expand and secure their access to existing oil and gas resources, and the logistical challenges are considerable. From the building of new pipelines, such as TANAP/TAP or South Stream/Turkish Stream, to the construction of huge ports for the landing of liquefied natural gas (LNG) that is transported over thousands of kilometres of increasingly unsafe international waters, the EU's need for fossil fuels force it to consider the complexities of international politics and national sovereignties (i.e. to face and overcome numerous geopolitical challenges). Related decisions, particularly those over investments in energy infrastructure have enormous geostrategic implications: for example, winning the race over the exploitation of new gas fields in the Caspian region is a precondition to the construction of new transport lines, just as finding and exploiting new deposits in the Norwegian Sea or Arctic Circle exposes the EU to Russia's territorial claims in the region. So far, the EU has performed surprisingly well. Notwithstanding the Nabucco controversy and the failure to reach acceptable transit terms with aspiring EU member Turkey, extensive energy diplomacy by the Commission and member-state governments led to the signing of a supply agreement with Azerbaijan in 2013, an improvement in communications with Russia over its gas supplies, and helped foster a fledgling international spot market for LNG.

Still, EU policymakers struggle with the need to strike a delicate balance between calls for change and the need for stability. The goal of reducing external energy dependence, or at least diversification of external suppliers and routes, is constrained by the politically risky nature of determining exactly how, when, and where energy security is to be realized. When one calculates the relative weakness of the third pillar of the EU's energy security plan, one must admit that comprehensive energy security is not only a moving target but also a very distant one. Ephemeral as real energy security may be, having it as a target – just like achieving an environmentally sustainable and largely carbon-free economy, and building a smoothly functioning and competitive internal energy market – provides a goal set that allows for a broad portfolio of activities which in the end may well lead to a more robust, if not quite autonomous, energy game plan.

Unfortunately, as sound as the logic of the three-pillars approach to comprehensive energy security may be, the EU did not have its key component in place, the functioning internal energy market, by the beginning of 2014. This is not to say that significant progress has not been made. Competition increased in the first years of the twenty-first century as member states implemented, at least in part, the so-called *Gas Directive and Electricity Regulation* of 2003 and the establishment of the unbundling regime (i.e. the 2009 Electricity and Gas Directives (EP/Council 2003, 2003c, 2009a, 2009b). Most European citizens can now choose between different electricity and gas providers. However, the 'unbundling' between generator/producer and service delivery has not yet been universally implemented across the Union. Very important for the consumer, prices of oil, gas, and electricity still vary significantly from one member state to another: for example, in the first semester of 2013, household consumers who cooked with or heated their water or homes with natural gas in Austria paid €0.767 per kWh (including taxes), while just over the border in Hungary, consumers paid only €0.432 for the same product and use; and electricity consumers in the United Kingdom pay €0.174 per kWh, whereas the same consumer in France paid just over €0.147 per kWh. Location always matters, but in the EU, political borders still seem to matter more than geography. Despite increasing its LNG imports substantially since 2002 (making up almost 20% of all EU gas imports in 2012) and reducing Russian gas' share of the import market (Russian gas accounted for only 23% of total EU gas supplies in 2012), dependence on Russian gas in several of the Central and Eastern member states (as well as Germany) remained high. Furthermore, Germany, the bastion of environmental austerity and cheerleader of sustainability, actually increased its consumption of coal for electricity generation between 2012 and 2013, simply because US coal had become so abundant and cheap as a result of its shale gas boom, illustrating the energy economics' equivalent of Mark Twain's aphorism, 'principles have no force, except when one is well-fed'.

The discussion so far demonstrates that security is the operative term in the EU's energy portfolio. The term is ubiquitous throughout the EU's energy-related legislation, from its directives and regulations on gas and electricity supply and distribution to the Commission's Communications on diversification, the increased use of renewables, or climate change. For most energy consumers, energy security translates into the stable supply and price of energy services required on a daily basis, specifically fuel for transportation and electricity and heat for homes, businesses, and industrial facilities. Energy is vital to economic and political stability, a fact that politicians ignore at their political peril. In Europe, where most states need to import primary resources such as oil and gas, failure to deliver such stability can risk government legitimacy in the eyes of their citizens and, either through unrest or electoral politics, endanger their hold on power. The security of supply and stability of price are thus central components of domestic politics and, as a logical extension, a central point of contention and concern for EU member states. Likewise, governments and businesses in supplier countries that are rich in energy resources, such as Russia, Norway, Algeria, and Qatar, require access to consumer markets, upstream investments, and price stability in order to secure and plan their deliveries.

The nuclear portfolio

Looking at the EU's three-pillar approach, one can easily identify key policy areas that are implicitly included or, for political expediency and Community organization, excluded. As the discussion above illustrates, for example, there is little mention of nuclear power, despite its continued important role. Nuclear power is a special case in the EU energy policy (Nuttall 2010), one that places it partially outside the box of the EU's broader energy portfolio while concurrently focusing the Commission's activities and efforts on issues of safety and non-proliferation. In principle, 'It is for each member state to decide whether to use nuclear power' (European Commission 2010c) and, according to the Commission (2006), decisions relating to nuclear energy 'can have very significant consequences on other member states' as relates to EU climate goals and dependence on imported energy. Thus, while the member states 'are neither obliged nor prevented from using nuclear energy' (Koranyi 2011: 195), the Commission tacitly backs member states in their choice to use nuclear energy, it just simply cannot afford to promote it, irrespective of how relevant it may be in meeting emissions' targets, maintaining reliable supplies of electricity, or balancing the EU's overall energy mix.

According to Eurostat's regularly updated online *Energy Trends* (2014), nuclear 'accounted for the highest share in primary energy production in EU-28 in 2012 (29 %)' well ahead of renewables (22 %), solid fuels (21 %), and gas (17 %). Given the obvious importance of nuclear power in its energy mix, one would expect a fairly high level of interest, even focus, on nuclear power as an alternative to fossil fuels for electricity generation. However, the Commission does not distinguish between nuclear and other sources in applying its regulatory framework for the single market in electricity (Nuttall and Newberry 2010). The difference is where completion and environmental sustainability play essential roles in shaping EU energy policy: in the nuclear domain the equivalent instruments were safety regulations on working conditions, hazardous materials, and non-proliferation (Housiadas and Dimitriou 2013).

The quiet and rather cautious position of the Commission on the nuclear portfolio is understandable. Nuclear energy is controversial: divisions – legal, practical, and most importantly political – prevent unity between the member states. Some member states, such as Ireland and Austria, adamantly avoid its use, while others, such as France and the UK, remain fully committed. Also, because the salience of electricity security varies across the EU, some member states, particularly the Central and Eastern European ones that are highly dependent on Russian energy imports, see greater benefit in adding nuclear power to their energy mixes than do others (Nuttall and Newberry 2010). Many of the states that oppose nuclear do so on political grounds (Helm 2012), or are already sufficiently diversified in their electricity generation sources as not to need it.

Nevertheless, the member states do agree on the need for extremely tight safety standards and the security of radioactive supplies. That has been and remains the focus of the Commission's attention. EU rules governing nuclear safety integration, harmonization, and use for energy production exist within the framework of the European Atomic Energy Community (Euratom), founded in 1957 as one of the two Rome Treaties (Chapter 3). The history of nuclear governance in Europe followed a parallel but separate path to the rest of the integration process. According to the Euratom treaty, the Community is supposed to 'Establish uniform safety standards to protect the health of workers and of the general public'. Building upon earlier secondary legislation on informing the public in the event of emergencies (Council 1989) and protecting outside workers who could be exposed to radiation (Council 1990d, 1997), several additional directives have been passed since the late 1990s, including: ones on safety standards and radiological protection of workers (Council 1996) and safety at nuclear installations (Council 2009a), which have strengthened the rules laid out in Article 30 of the Euratom treaty; the *Waste Directive* (EP/Council 2008a), obliging member states to deliver national plans to the Commission (the first by August 2015); and the more general transportation of spent fuels and waste and other dangerous goods (Council 2006). The two latter directives followed the passing in 2003 of a related directive on the management of waste (Council 2003) and a Commission recommendation on nuclear plant safety (European Commission 2004a; see also Council 2003a), which stemmed from a 2002 Commission Communication (European Commission 2002). The most important of these is by far the *Nuclear Safety Directive*, which for the first time established European-wide requirements and made IAEA safety fundamentals binding on the member states, but most importantly left the implementation to the member states, which at least in part explains the limits of EU nuclear policy. Several directives were repealed and replaced in 2013 when the Council issued a new directive on safety standards that established 'uniform basic safety standards for the protection of the health of individuals subject to occupational, medical and public exposures against the dangers arising from ionising radiation' (Council 2013a).

In practice, responsibility for the safety of nuclear installations, as well as the systems of licensing and supervision, lie with the member states, and nuclear plant operators (i.e. licence holders) bear responsibility for disasters. Member-state reporting on implementation only began in 2014 and should continue every three years hence (Housiadas and Dimitriou 2013). In the wake of the 2011 Fukushima nuclear accident in Japan, the Commission cooperated with the European Nuclear Safety Regulators' Group (ENSREG), an independent body of senior officials from related nuclear and energy regulatory bodies, to carry out voluntary stress tests. Those tests revealed that there was room for improvement and that there were 'significant differences in the national approaches' to nuclear safety (European Commission 2013h). In October 2013, the Commission moved to change the existing EU framework on nuclear safety by proposing a *New Nuclear Security Directive*. Their proposal included enhancing the regulations to the latest technical standards, giving national authorities greater independence from their governments, allowing for the setting of EU-wide safety objectives, and establishing a system of peer reviews to verify compliance (run through the Commission). Two additional measures included opening up the industry to public scrutiny and requiring public disclosures by licence holders and regulators (i.e. creating the opportunity for citizens to participate in the licensing process) (European Commission 2004, 2013i).

The EU formally amended its *Nuclear Safety Directive* (Council 2014a) in July 2014, and full implementation (transposed into national legislation) should occur by 2017. The final consolidated version of the new directive included most of the Commission's proposals, specifically: strengthening the independence of national regulatory authorities from their national governments; implementing the Commission's proposal for a system of peer reviews of national safety assessments, the findings of which will be made public; requiring safety re-evaluations of all nuclear power plants no less than once every 10 years; and formalized requirements that operators release information to the public (both in normal and crisis times).

The Commission also collects data on the member states' nuclear installations and practices as part of its oversight role. For example, a 2009 Commission recommendation (European Commission 2009d) impels users and holders of nuclear materials to keep records on the flow, processes, and stocks of their materials, and declare them to the Commission. Inspectors under the employ of the Commission travel to all operating reactors, collect data, and publish annual safety reports. On the external front, DG Energy communicates regularly with nuclear authorities around the world, promotes global standards, and participates in related forums, such as the European Nuclear Energy Forum (ENEF), a platform for discussions between stakeholders in the nuclear field. Yet, almost all of these activities, internal and external, revolve around safety issues.

For the EU, the nuclear portfolio is a matter of how to go about it rather than whether to do so. This is quite explicitly revealed in the Commission's Communications on nuclear policy. Officially, the EU supports a framework for nuclear energy in those member states that choose to use it, as long as they do so 'in conformity with the highest standards of safety, security and non-proliferation as required by the Euratom Treaty' (European Commission 2007b). Nevertheless, it is clear that there is support for nuclear energy as part of the EU's overall energy mix, particularly in light of the climate change issue. It is particularly telling that in the Commission's 2007 Communication, *An Energy Policy for Europe*, it dedicated less than 600 words to the nuclear question and chose to describe nuclear in a rather positive light as 'one of the cheapest sources of low carbon energy', 'less vulnerable to fuel price changes than coal or gas', and 'one of the ways of limiting CO_2 emissions' (European Commission 2007). Still, recognizing nuclear energy's benefits for the larger goal of building a low-carbon economy is a far cry from actually placing a stamp of approval on the matter or accounting for its economic sustainability, which as Helm (2012) points out, is questionable. Even the most visible EU-backed project in the nuclear field, the International Thermonuclear Experimental Reactor (ITER) is a distant possibility. Given its costs to the EU (over €1 billion between 2012 and 2013) and its very long timeline, it is reasonable to ask whether or not EU funds could be better spent on infrastructure for its fledgling internal energy market.

In addition to all these points, nuclear has clear military (France and the UK operate nuclear weapons programmes) and profoundly serious health implications, which separate it policy-wise from other primary energy fuels. The military dimension is clearly a subject and prerogative of the two member states involved, an 'unquestioned core of the defence postures of both France and the United Kingdom' (Jasper and Portela 2010), and there is no serious discussions at the EU level of integrating member-state nuclear forces in any future plans for coordinated or common defence forces under the rubric of the EU's evolving *Common Security and Defence Policy* (CSDP). In the area of health and safety, on the other hand, there has been a significant amount of cooperation between the member states for many years, particularly in the area of safety and emergency response. Thus, in addition to the legal framework in place, 'there is a "non-binding acquis" that is built on fundamental common principles' (de Esteban 2002).

This was not always the case. Nuclear installation safety and waste disposal were not particularly important to the original signatories of Euratom, and for much of the history of the EU well into the 1970s, there was virtually 'no Community activity directly dealing with nuclear installation safety' (de Esteban 2002). By the time safety became an issue (following the Three Mile Island incident in the US and later the Chernobyl disaster), Europe's nuclear industries were prospering economically, but following very different paths with different set-ups and regulation schemes in the respective member states. Since then, harmonization was made easier merely by the fact that the treaty basis (Euratom) for such cooperation had long been in place, if unused. Nevertheless, the nuclear industry in Europe faced, and continues to face, serious obstacles, the most important of which being the combined effect of public concern and national politics.

There simply is no consensus on the use or future of nuclear power in Europe by the member states, and certainly not by the population. As Dieter Helm (2012: 124) observed, opposition to nuclear power is 'an ideological position for most; a matter of faith rather than a question open to debate and discussion'. What is more, the member states themselves have changed their positions on nuclear power over time. In Sweden, for example, a 1980 referendum called for a phase-out of nuclear power, a decision which, though never fully implemented, was reversed 30 years later. In Germany, once one of the biggest supporters of nuclear power in Europe, Chancellor Gerhard Schroeder's government, which included the Green Party, decided in 2001 to forgo all new construction and close all of its nuclear power plants by 2020. Although Angela Merkel's government later took a less harsh line at first, it moved swiftly after the Fukushima disaster to close eight power plants, promising the decommissioning of its nuclear facilities within a decade (Helm 2012). Meanwhile, Finland and France are expanding their use of nuclear energy, while the Netherlands, Poland, Sweden, the Czech Republic, Lithuania, Estonia, Latvia, Slovakia, the UK, Bulgaria, and Romania are each actively debating their respective nuclear energy policies. Even if the Commission were to fully engage itself and take sides in the nuclear debate, it would have to overcome not only the opposition of some of the member states but also face off against a public that is extremely sceptical about nuclear power (European Commission 2010c).

Finally, as climate change became more pressing of an issue, hopes for a nuclear renaissance rose. Indeed, Europe's growing concerns about climate change should have been 'heaven-sent for the nuclear industry' (Helm 2012: 121): after all, nuclear energy generates low-carbon electricity without producing much CO_2 and, historically, was relatively cheap in comparison to early wind and solar power. However, the Fukushima disaster brought back old fears, and the US shale gas revolution in the mid-2010s 'pulled the economic rug from under nuclear's feet' (Helm: 21). Therefore, although there were 132 nuclear power plants operating in 14 member states in June 2013, nuclear power was as controversial as ever in January 2014.

Parallels and paradoxes

There is yet another, more critical way to understand the EU's approach to comprehensive energy security and, thus, the components of its energy portfolio. Whereas the three-pillars approach aptly describes the EU's energy portfolio as the Commission frames it, one can also observe the EU taking three parallel, and sometimes contradictory, approaches to achieve energy security (Pollak et al. 2010) that focus on reducing demand internally, expanding its supply options, and exporting rules externally. First, it is actively pursuing the reduction of overall energy demand by increasing efficiency (also known as reducing energy intensity) and the share of renewable fuels in its energy mix. Demand reduction, in turn, serves at least two other purposes. Hypothetically, this should reduce the need to import fossil fuels from outside the Union, hence mitigating the negative geopolitical consequences of external dependence by shrinking the security-of-supply issue to the internal matter of the (un)reliability of renewable sources of electricity. The end result of this approach is that it also reduces greenhouse gas emissions and, thus, enhances environmental sustainability. The focus of these actions matches elements of the internal and multidimensional rubrics of our energy policy typology. Second, the EU is supporting private commercial projects to build new pipelines and LNG terminals that increase natural gas-import capacity, diversify import routes, and expand the roster of natural gas suppliers. Combined with a reduction in demand, new sources of traditional fossil fuels enhance the stability of supplies flowing to Europe. The downside of that approach, however, is that it may also reduce the impetus to invest in alternative or substitute fuels, by keeping fossil fuel prices low enough to prevent alternatives from becoming competitive. The only way to overcome this dilemma is to support the price of alternative fuels until the scale of their production is such that they can compete with oil, gas, and coal. Unfortunately, while such an approach meets the requirements to transform Europe into a less carbon-intensive economy, it also flies in the face of decades of liberalization, privatization, and efforts to eliminate subsidies and other unfair market practices. This policy set, as one can see, is complex and fits all three rubrics of our energy policy typology. Third, the EU is using diplomacy and rule export to incorporate bordering regions into a standardized set of operating rules and regulations. In other words, the EU is exporting European standards of governance in the energy sector, and is doing so with quite some success. The best example of this is the Energy Community, an international organization established by treaty in 2005 (in force since 2006) that brings the EU together with its regionally close suppliers and partners from South-East Europe and the Black Sea region (as well as key global partners) in an attempt to standardize rules and protect energy flows to Europe while securing its foreign, direct investments in the energy sectors of its supplier states. This clearly falls under the external dimension of EU energy policy.

These approaches cannot be understood merely as separate pillars as such because each operates concurrently with varying effects on the other. On the demand side, the EU has set some mighty goals, some of which are binding. Its plans to raise the share of renewables in its total primary energy supply, including raising the share of renewable energy (use of biofuels, hydrogen, etc.) in the transport sector by 2020 (EP/Council 2009d), and its ever-increasing emission reduction targets point to an active approach to fundamentally change how energy is consumed in the EU. In January 2014, the Commission added new objectives for 2030, including reducing emissions to 40% below 1990 levels, increasing the share of renewables in overall consumption to 27% (up from 13% in 2012), and strengthening the EU-ETS market. For the latter, it proposed adding a market stability reserve in 2021 that would adjust the supply of 'auctioned allowances downwards or upwards based on a pre-defined set of rules and would improve resilience to market shocks and enhance market stability' (European Commission 2014: 8).

Yet the Commission backtracked between 2003 and 2009 on biofuels as it became clear that some fuel products could have substantial negative impacts on biodiversity and land use, leaving many investors and supporters of alternative fuels in the lurch. Similar retreats occurred in the German biodiesel market when the German government pulled the plug on subsidies, and when the

Spanish government arbitrarily changed the feed-in tariffs for small-scale photovoltaic power generation. Despite such backtracking, the EU's overall approach of setting targets for reduced emissions has been good for Europe's environment and its long-term security. The bad news is that those same targets may prove to be a disaster for Europe's shorter medium-term energy security, because it sometimes moves against the natural flow of the market and inflates energy prices.

In two of the most significant energy sectors, electricity and heating, reducing CO_2 emissions has its largest impact on coal-generated power and heat. Coal is by far the most CO_2-intensive of all the primary energy sources. Since member states retain sovereignty over their own energy mix, they must come to an agreement about which of them are going to switch from coal to alternatives and at what rate, if the member states are to reach their collective goals. However, the market generally drives decisions over specific national energy mixes and, in Europe, it is cheaper to generate electricity with coal than with gas. According to energy market observer Platts (2013), for example, the profitability gap between coal and gas in Germany 'widened to Eur26/MWh based on year-ahead contracts for power, coal, gas and carbon emissions' in 2013. As a result, Germany, the champion of the shift away from coal, started using more of it; in large part because US coal surpluses hit the market as a result of its development of shale gas resources. And it is the US, the previous champion of coal, that was actually reducing its emissions at a faster rate because its newly added and cheap shale gas surpluses were increasingly being used in its energy mix. The fuel mix issue remains a problem for Europe, but it is one that member states seem least likely to agree upon.

Nevertheless, discussions over which fuels will be used by the Union remain a necessary, if rather controversial, policy area for the Commission. Renewables such as wind, solar, geothermal, and hydropower are the most obvious alternatives to coal for electricity and heat generation. Europe's renewable share may even eventually climb beyond the 20% level with technological advances and some hefty investment. However, given the current state of technology, the efficiency of conversion processes, and the availability of reliable renewable options, most of the savings up to 2020 will not come from an increased use of renewables.

Reducing reliance on coal and oil and switching to natural gas for electricity and heat generation is a key component of EU strategy to reduce its carbon footprint. To meet its goals, the EU will, therefore, have to increase its foreign dependency on this resource. In essence, the EU is trying to win energy security in the long run by significantly degrading it in the short run, like sprinting to win a marathon. Given this dilemma, the EU has taken specific steps to enhance its security. A Council directive (EP/Council 2004a) created a legal framework to safeguard gas supplies in the case of supply disruptions by establishing a Gas Coordination Group to liaise between the Commission, the member states, industry, and consumers. In 2006, the European Council endorsed the Network of Energy Security Correspondents to provide an early warning system for potential threats to the EU's energy supplies. In November 2008, a Commission Communication (European Commission 2008c), *Second Strategic Energy Review – An EU energy security and solidarity action plan*, identified key infrastructure projects targeted for support, including the southern gas corridor, LNG import and storage facilities to secure a diverse LNG supply for Europe, new interconnections to the Baltic region, the Mediterranean Energy Ring, and key north–south gas interconnections within Central and South-East Europe. This was rounded off by a 2010 regulation that required the competent authorities in each member state to conduct risk assessments, and establish both a Preventive Action Plan and an Emergency Plan to guarantee that all 'necessary measures are being taken to ensure [...] continuous [gas] supply, particularly in case of difficult climatic conditions and in the event of disruption', and to prepare 'measures to be taken to remove or mitigate the impact of a gas supply disruption' (EP/Council 2010a).

The EU's dual, and sometimes contradictory, goals of decarbonizing its economy and enhancing its energy security has been made all the more difficult to achieve in light of the US shale gas boom, as illustrated by the German example. Some member states, such as the United Kingdom and Poland, are hoping to follow the US lead. Although the Commission deems the EU's shale gas reserves 'to be significant' according to a January 2014 Communication on the subject, so far the results have been poor and controversial, as many Europeans oppose those efforts and fear the negative environmental consequences of the water–intensive, horizontal-drilling method known as hydraulic fracturing, or fracking. As of January 2014, the technique was legal and being deployed in Denmark, the Netherlands, Poland, and Romania. It was also legal in Ireland, but no wells were yet developed, and it was de jure legal in Germany but de facto not, as Berlin imposed a moratorium in 2013 until further research could be undertaken. The Commission issued recommendations in January 2014 (European Commission 2014b) accompanied by a Communication on 'the exploration and production of hydrocarbons (such as shale gas) using high volume hydraulic fracturing in the EU', in which it recognized that 'shale gas can be a possible substitute for more carbon-intensive fossil fuels, an indigenous source of natural gas reducing dependency on non-EU energy suppliers, as well as a possible driver of jobs, economic growth and additional source of public revenues' (European Commission 2014a).

The debates over nuclear power and hydraulic fracturing (fracking) in the EU are good examples of the difficult policy choices faced by the Commission. On one hand, the Commission has championed environmental protection; on the other, it cannot ignore the possibility that Europe may be able to replace external gas supplies with indigenous fossil fuel resources, or substantially reduce its carbon footprint and external energy dependence on gas through increased use of nuclear power. While the former is a battle that is just beginning and will increasingly find its way into the EU's energy portfolio, the latter has oscillated between a panacea and political hot potato.

Concluding remarks

EU Energy policy is not easy to define. European Union Energy Commissioner Andris Piebalgs broadly defined it in 2007 in economic terms as 'a common energy market that works under particular investment conditions' (Euractiv 2007). The difficulty is not limited to EU politics either. While some consider it 'too complicated to be left to the politicians' (Frank 2006), former US President William Jefferson Clinton called energy policy 'the key to national security', 'fighting global warming', and 'more jobs and higher incomes'. (CNN 2008). Whichever way one categorizes energy policy and whichever issues one includes, one thing is certain, there is precious little consensus over precisely what issues do or do not fall into an energy policy portfolio.

This is because energy issues are not mere matters of resources, services, or the environment; energy itself is a political commodity. When it is available and cheap, economies grow rapidly (as was the case in the post-war boom years in Europe) and the price of food and mobility are low. When supplies run out, are interrupted, or prices spike, the economic fallout on households and businesses can be disastrous. Therefore, no policymaker who cherishes his or her job can afford to ignore energy; it is a proverbial bread-and-butter issue. As any student of political science will attest, policymaking is an imperfect and complex process that involves many actors and forces, each driven by different, competing, and often contradictory interests. Both the array of issues and the constraints on policymakers increase when one considers not a single national energy policy, but rather a conglomerate set of national policies organized and, to an increasing extent, coordinated at a supranational level.

This chapter began by organizing energy policies into a typology of three dimensions, an internal, an external, and a multidimensional. Among the most notable policies that fit into these dimensions are: the establishment of a functioning internal energy market and all that comes with it (liberalization, deregulation and re-regulation, subsidies, integrating network infrastructures, and harmonizing taxes and prices); continuing debates over controversial new technologies (fracking) and old ones (nuclear power); reducing external dependency (through diversification of suppliers, building new pipelines, energy diplomacy and rule export, and reduced demand through increased efficiency); and finally, making energy use environmentally sustainable by addressing vital reductions in energy-related greenhouse gas emissions. Throughout the course of the chapter we show how each of these issues fall within the EU's energy portfolio, which the Commission organizes into three pillars: sustainability, competitiveness, and security of supply.

The EU wants to move to a new industrial age, a new green economy that promises economic growth, jobs, and a clean environment, but it faces very difficult challenges. In order to implement its three-pillar strategy, the EU is taking concerted, but limited, action. However, the goals it has and the approaches it is taking are in many ways implicitly contradictory. It is difficult to guarantee fair competition in the energy sector: suppliers are unevenly distributed in location and number, and companies need long-term guarantees to justify expanding their infrastructure, neither of which fits well with a competitive market or allowing third-party access to a company's infrastructure. It seems obvious that the best way to decarbonize Europe's economy is to reduce coal consumption for electricity generation and petroleum for transportation. That explains in part why, in a 2011 Transport White Paper, the

Commission set a goal of reducing emissions in the transport sector by 60% from 1990 levels by 2050 and by around 20% from 2008 levels by 2030 (European Commission 2014: 14). However, replacing either too quickly only increases prices and stress on consumers. Indeed, pairing the aims of competitiveness with its climate goals often seems to be a Sisyphean task.

Reducing the EU's external energy dependence is another problem fraught with contradictions. Although the Central and Eastern parts of Europe are overly dependent on Russia for natural gas, Russia is just as dependent on Europe for its investments and technologies, especially in the LNG sector. Meanwhile, developing new indigenous gas resources, which would not only help reduce foreign dependence but also create jobs and help expedite decarbonizing the economy, is opposed by many for its very real negative environmental consequences. Finally, by decarbonizing the economy, and raising the costs associated with industrial production, Europe is in many ways not only decarbonizing but also implicitly deindustrializing, pushing traditional economic opportunities for manufacturing and labour out of Europe to places where environmental conditions and emissions' controls are weaker.

Meeting together in Messina, Italy in 1955, the six founding members of the EEC declared that '[p]utting more abundant energy at a cheaper price at the disposal of the European economies constitutes a fundamental element of economic progress' and all 'arrangements should be made to develop sufficient exchanges of gas and electric power capable of increasing the profitability of investments and reducing the supply costs' (Messina Declaration 1955). The EU has come a long way since Messina and its energy portfolio has grown substantially. At one point or another, the EU has taken up each of the issues described in this chapter (and many more), in some cases resolving matters, and where not, setting bold goals and targets, many of which continue to elude EU policymakers.

Before exploring in detail the changing nature of EU policy and its theoretical underpinnings, we need to review some of the fundamental principles of energy economics, particularly its end uses and how the global energy balance shapes and conditions policy choices.

Chapter 2

Principles of Energy, End Uses, and the Global Energy Balance

This chapter introduces readers who do not have a background in energy to the key information that they need to make sense of this policy area. It begins by explaining some of the fundamentals of the economics of energy, thereby building upon the foundational principles of energy laid out in the Introduction – that energy comes in two forms, kinetic or potential; that energy can be converted from one to another; and that a variety of energy-carrying resources store energy unequally and are either finite or not. With a firm grasp of energy's unique and distinguishing economic concepts in hand, we proceed to examine its three main end uses: electricity, heating, and transportation. Keeping in mind the definitive goal of comprehensive energy security, we identify how these end uses drive policymakers to incorporate internal, external, and multidimensional policies. We conclude by contextualizing these end uses in terms of the global balance of supply and demand for the most widely used finite energy resources (oil, gas, coal, and uranium).

Basic principles of energy economics

Appreciating how energy is used requires at least a basic understanding of the terminology, actors, markets, and forces at play, or as in politics, the who, what, when, and where of the energy business. Energy concepts are not easy and analysts tend to thrive on special terminology that often leads the layman to misunderstanding, in many cases significantly, what is being referred to.

Primary energy production refers to any kind of extraction of energy products from natural sources into a usable form. Primary production occurs when natural sources are exploited, no matter whether that source is a coal mine, crude oil field, hydropower plant, wind farm, solar array, or the fabrication of biofuels.

Importantly, the conversion/transformation of energy from one source to another, like the creation of electricity in a thermal power plant, is not referred to as primary energy production. *Energy consumption* refers to energy delivered to the final consumer for use. It excludes losses and deliveries for the purpose of transformation or use by the energy industry. The term *generation* is used almost exclusively in reference to electricity and it includes the total power output of electric power plants. All power plants require some input source, often called a *fuel* or *feedstock*. A fuel is any material that is burned or altered in order to obtain energy. *Fossil fuels* are fuels resulting from the compression of ancient plant and animal life formed over millions of years. The most well-known fossil fuels are oil, coal, and natural gas, which taken together constitute the bulk of energy feedstocks. A feedstock is any raw material used in an industrial process to create something else. Petroleum is a feedstock used in a refinery to produce everything from asphalt to gasoline. Finally, imports and exports of energy always exclude transit, that is, resources that pass across one country on their way to another.

Derived demand

The economics of energy is largely a product of *derived demand* driven by the desire to use energy services, independent of the source. When the average European citizen turns on the television, whether to watch the news, a soap opera, or the Euro Cup, he or she rarely knows with certainty what the original source of the energy was. All that really matters is whether there is electricity. Even if he or she did know the original source of the electricity, or its impact on the environment, it is unlikely that approving or disapproving of it would dissuade a hungry person from turning on an electric stove to cook a meal. The most important economic issue for the consumer is, therefore, not about which energy commodity generates the motion of their car or electricity in their home, but rather which energy service delivers the means to conduct such activities as efficiently (i.e. cost effectively) as possible. This source-independent nature of energy consumption is essential to understanding consumer energy demand.

Gasoline is a good example. Consumers buy gasoline to drive cars. The amount of gasoline they buy is in direct relation to how much they drive, which is inversely proportionate to their vehicle's fuel efficiency. If they need to drive 500 kilometres, drivers will fill the tank. If it is too expensive to do so, they will not drive that distance. For some time, gasoline has been the least expensive (that is, most efficient) way to provide mechanical energy to a vehicle. If an alternative source were to offer higher efficiency, that is, a lower cost per kilometre, then gasoline would lose its allure. Thus, demand for gasoline is derived from consumer choices about the distances they plan to drive in their vehicles (Sweeney 2001).

Electricity is no different. Consumers use electricity regardless of its origin. They pay for the service because it powers their basic home needs and they pay rates based on the efficiency of the products they use. An old refrigerator consumes more electricity than a new one per minute of use. Thus, the cost of energy to the consumer drives choices about energy-consuming products from cars to light bulbs. As a result, the behaviour of energy consumers is a factor of efficiency and is largely determined by price stability. Volatile fluctuations in prices directly affect efficiency, with upward swings in the former inversely affecting the latter.

Distinguishing between resources, costs, and end uses

The most common method to distinguish between the ways in which energy is consumed is to differentiate between markets for oil, gas, coal, and power (electricity). This allows observers to calculate existing and projected supply and demand for each of the primary energy sources, plus electricity, including imports, exports, and transformation. Many of the International Energy Agency's statistics are arranged in this manner. Parallel to this source-based division, energy studies tend to distinguish between end use by sector and product groups. These include services and households (solid fuels, gas oil, natural gas, renewables, electricity, and heat), industry (solid fuels, petroleum products, natural gas, renewables, and electricity), and transport (motor spirit, diesel oil, jet fuel, and electricity). This is the method used by the European Commission. For the purposes of this chapter we use a simplified hybrid of these two methods, focusing on and distinguishing between the end uses of energy: electricity (power), heating, and transportation. Understood in the overall context of an energy policy typology, this approach allows us to match consumer and national security needs with policies aimed at achieving comprehensive energy security (see Chapter 1).

As explained earlier, all energy is convertible, and electricity is a converted energy source. Therefore, all energy resources, from flowing water, to nuclear, to coal, and even horsepower, can be used to generate electricity. This is not true, however, for the fuels that drive our transport industry (with the notable exception of electric trains). Transport fuels are derived predominantly from hydrocarbon-rich energy resources, such as oil, coal, natural gas, and more recently, certain biofuels such as ethanol. Moreover, since some industries want to generate their own electricity at different power levels for their particular machinery, hydrocarbon fuels may be purchased to run local generators. Therefore, the very first issue to take up when formulating energy policies has to be a clarification of which sector a policy applies to. Building nuclear reactors or wind farms will increase the electricity supply, but will not affect the price of gasoline at the pump. This is a good example of how some policies in the internal dimension have little external impact, despite their appeal or cost.

The price of gasoline at the pump is dependent upon, but not equal to, the price of petroleum. This is because gasoline is a product refined from crude oil, which is usually mixed with other chemicals and components, including biofuels, and its market is more limited than that of oil in general. The gasoline versus oil distinction illustrates the fine, but important, difference between the economic value of an energy resource and that of an energy commodity. Energy resources occur naturally. They include everything from wind, sunlight, and the water in our oceans to the geothermal heat and unknown quantities of fossil fuels buried beneath the earth's surface. Energy resources that are mined, drilled, or captured are converted into commodities such as gasoline, heating oil, and electricity in order to conduct tasks designed to make our lives easier (see Sweeney 2001: 4513). Energy commodities have a commercial value and they are traded openly on international markets. Simply stated, energy resources are used to produce energy commodities, which in turn are sold and traded for use in providing services. Petroleum, natural gas, and electricity are all heavily traded commodities. Wind and water, for the moment, are not.

Embedded in the final price of energy services are the costs involved in discovering new resources and delivering them to the market as viable commodities, fuels, and services. The consumer may drive demand and thus largely determine the price of an energy resource on the open market, but a myriad of activities along the way significantly affect the minimum end-use price. Among these are the capital-intensive costs of resource discovery, extraction, refinement, and conversion. In the oil industry, the business of exploring, drilling, and pumping oil is called the upstream, and the costs and risks of failure are enormous. A series of midstream processes including shipping and piping deliver oil and natural gas to the downstream phase, which includes refining and retail sales. While the discovery of a new large deposit of petroleum may increase the long-term supply outlook, the turnaround time from discovery to commercial viability is usually more than a decade. Moreover, while the upstream is subject to the risk of not discovering a resource, adding calculated risk costs to the price of oil and natural gas, the mid- and downstream phases are subject to everything from bad weather to political upheaval.

Sitting atop a valuable resource alone does not make a country rich, but being able to convert it into a commodity may. Canada sits on top of one of the world's largest oil reserves. Unfortunately, that oil is not in liquid form, but rather is a thick tar-like substance. Thick and uneasy to handle, it is nevertheless oil and in order to convert it into liquid form – useable by a refinery to produce gasoline and other products – developers of Canada's oil fields have to pump in huge amounts of water and heat, dramatically raising the cost of extraction. When oil prices are high, this makes economic sense. When prices drop, it can spell financial ruin for all those involved.

Embedded costs are a fundamental aspect of the delivery of energy services. Building a power plant is a major investment. Integrated into the final price of electricity delivered to end users are the capital costs of building the power plant and connecting it to the grid, running costs such as operation and maintenance, and financing costs to repay the original capital. If a country is rich in wind energy, such as when they have long coastlines or windswept plains, someone has to build a large number of turbines, and deliver the electricity to the grid. Along the way, all these costs are reflected in the final consumer price. It is no different for gasoline, which is a product refined from oil. Because the complete energy chain usually involves hefty foreign investments, government oversight, regulations, and negative environmental consequences, one can observe all three dimensions of energy policy in everything from the price of a single litre of gasoline to the cost of running a computer.

Storage, predictability, and negative externalities

Added to the commercial viability and economic value of any energy resource is the degree to which it is predictably available. Reliability is a core component of comprehensive energy security. The best way to predict availability is when one can store a resource and easily access it when needed. Most fossil fuel resources such as petroleum, coal, and natural gas are storable. Among renewables, only water and biomass fuels can be stored en masse. Other major renewable energy sources such as wind and radiant energy cannot, although newer solar-thermal systems use sunlight to create heat and store the heat for later use. When a resource is storable and regularly used from that source, such as in the cases of a water reservoir or oil bunker, then the rate at which that resource is used limits how much is available later. Storable resources are thus predictable, but must be replenished upon use. Seen in economic terms, they operate under the rules of a zero-sum game. Likewise, non-storable energy resources such as wind, solar, and running river water may be subject to natural disruptions in supply, and hence are considered less reliable. Therefore, when analysing the efficiency of an energy resource, it is important to keep in mind not only supply but also storability, and thus predictability.

Electricity, for example, is very difficult to store, particularly in large quantities. In fact, electrical energy must be consumed immediately or converted to chemical energy and stored in a battery; otherwise, it is dissipated (lost) as heat. Most electricity is consumed as it is produced; being converted back into other forms for different tasks, such as to thermal energy in the case of water heating, or first chemical, then thermal, then radiant when powering a light bulb. Policies that target the power sector always have to prioritize reliability of the power supply, greatly complicating choices over fossil versus renewable sources.

One very important feature of energy production is the hidden cost of negative externalities. In simple terms, a negative externality is the bad effect of actions taken by one set of parties on others (or everyone else), who neither had a choice, nor whose interests were taken into account. The negative externalities of energy production include, among others, air and noise pollution, global warming, and shifts in water flow. Although it is difficult to calculate the per capita cost of an externality such as global warming, the costs of responding to it are visibly increasing in terms of environmental clean-up operations, industrial conversion, and public health.

Because such problems are beyond the scope of any company, if not even countries, addressing externalities such as these is predominantly the domain of policymakers. Indeed, the discovery of global warming patterns has led to increased calls to rein in the mass consumption of energy through greater efficiency standards and underscored the importance to find alternatives. Intriguingly, the externality issue has also led to the establishment of an entirely new market for traders, namely allotments of permissible carbon dioxide emissions. Global negative externalities and their political effect correspond to the multidimensional aspect of our aforementioned energy policy typology (see Chapter 1).

When all the cost factors are calculated, from supply and demand imbalances to embedded costs and externalities, they can drive the price of an energy service to a level where it becomes prohibitive, forcing consumers to look for alternatives. In energy economics, those alternatives are called substitutes.

Conversion, exhaustion, and substitution

Analysts of energy policy should always remember that energy is highly flexible. Homes can be heated with electricity, but they can also be heated with natural gas, oil, wood, or even coal or solar power. Thus, energy commodities are technically substitutable for one another assuming one ignores negative externalities such as environmental damage. Since energy commodities are exchangeable, the price of one particular energy commodity is a function of other energy commodities and their price and cost of conversion.

Conversion technologies are the primary bottleneck in energy commodity substitution. A gas furnace is built to burn natural gas. It will not create heat with the input of wood or electricity. Traditional gasoline-driven automobiles will not run on alcohol, hydrogen, or pure electricity. Therefore, in order to realize the potential of energy commodity substitution, new technologies and the infrastructure to support them are required (internal dimension). The time and investment it takes to generate these conversion or multi-input technologies are significant obstacles to their implementation, and one of the most significant issues in the energy policy domain. Thus, when one speaks of energy independence, it is unrealistic to predict that it could be achieved in a time period less than it takes to develop the new technology, test it, commercialize it, pass it through the regulatory framework, and get it purchased by a large enough consumer base.

Substitution is an important factor in the economics of energy, and essential to any plan to reduce dependence on external depletable resources such as petroleum. According to depletable resource theory, originally posited by Lewis Grey in 1914, market prices of a given resource increase gradually over time until they reach an equitable cost of producing substitutes. That ceiling is reached, however, only as the resource nears depletion. In the parlance of the current public debate on oil, the signal of this ceiling is called 'peak oil', and there is debate and uncertainty about whether peak oil has been or will be reached anytime soon.

The problem of identifying the trigger to bring about the substitution of a finite resource lays in a calculation of confirmed supplies known as reserves, less current production rates that eat away at the reserves, with an added tally for the total amount of new discoveries. In the case of oil, this is illustrated in a reserves-to-production ratio, which calculates the number of years of oil remaining, assuming no new discoveries are made, and production is held steady. In 2012, the energy giant BP listed the R/P ratio for oil at 52.9 years, which was up from 40.5 years in 2006, 41 years in 1996, and 39.8 in 1986 (BP 1987, 2007, 2013). For comparative reference, the R/P ratios in 2012 for gas and coal were 55.7 and 109 years, respectively.

What is particularly intriguing with the R/P ratio is that the long-term outlook for petroleum supplies has remained relatively stable over the last three decades. The reason for this is that new deposits are found annually and there is still a lot of oil suspected, but undiscovered. For example, heavier oils, which are believed to exist in massive deposits in the tar sands of Canada and shale deposits of Colorado in the United States, as well as a myriad of deep-sea sources from Brazil to the Arctic, have yet to be fully explored, let alone tapped. Even if found, such resources are very expensive to prospect and come with significant negative externalities on the environment. Nevertheless, if global demand for oil rises and the long-term outlook for supplies decline, such resources can be considered as possible substitutes to the easy-access light petroleum that now makes up most of the production, trade, and consumption of oil.

Heavy oils that are more expensive to pump provide an excellent example of the substitution concept. For one, most of the conversion technologies required for its commercial viability are deployed upstream. The downstream consumer would still buy gasoline with little or no knowledge of its original source. This would also be true for coal-to-gasoline conversion technologies. What downstream consumers would face, however, is increased negative externalities from upstream production. Similarly, converting automobiles to use natural gas or biofuels, a process already underway, comes with its own set of negative externalities that have yet to be fully calculated. Moving entirely from the internal combustion engine to electricity may be the most definitive example of substitution, but as discussed earlier, battery storage and the infrastructure needed to charge the vehicles are bottlenecks that have yet to be fully resolved.

Nevertheless, substitution as seen from the energy economics' perspective is inevitable. It happened previously when oil replaced coal. Moreover, the specific energy services currently available to the market – fuels for vehicles, heating, and power – are not irreplaceable, or in economic parlance, energy services are not essential goods, those in which demand remains strong no matter how high the price becomes (see Sweeney 2001). Companies that provide services deemed essential to the public and are regulated to guarantee that they provide equal access to the service are called public utilities. Public utilities deliver such services as reliable water supplies, electricity, and in some cases, heating. There is little question that access to water is an essential service, but is this also true for energy? One can easily substitute one energy form for another, if the price is right. Solar power can heat and electrify homes; people can travel by foot or pedal power. Wood and other biomasses can be used for fire, which in turn can generate heat used for cooking. It is possible, if wanted, to survive with fewer energy supplies and less consumption. If gasoline were too expensive, it would become price-inefficient, and demand would tend to decline. For an essential good, such as drinking water or breathable air, it is difficult to imagine consumers willingly choosing to spend their disposable incomes on vehicle fuel and electricity rather than water or air, particularly if they had to pay for it and it amounted to their total disposable income. Likewise, there are complements to substitutable goods that are dependent upon the price and availability of other goods, rendering the latter essential to the former. For instance, certain vehicles require certain fuels, but if that particular fuel becomes prohibitively expensive over a long enough period, sales of the dependent vehicles will also decline.

There is little doubt that fossil energy sources, by their very nature, will eventually be exhausted, although there is great debate about when. At some point, through either uncontrolled market forces or coordinated public policies, conversion technologies will improve, and renewable energy sources will take up the lion's share of total prime energy consumption, just as happened two centuries ago. It is not a question of if, but rather when and how. At some point, the cost of depletable resources and the externalities of their use will price themselves out of the market and consumers will actively seek some form of substitute to satisfy their energy demands. In pursuit of comprehensive energy security, particularly sustainability and security of supply, governments can and often do try to expedite this process through a myriad of proactive and constraining activities, most notably: investments in research and development; increased or reduced taxes; tax credits and regulations; and other policy areas that squarely fall into the internal dimension but have substantial impact on the external dimension, while concurrently steering the market away from liberal principles by favouring one energy source over another.

The changing marketplace

Finally, the economics of energy has been greatly complicated by significant, sometimes sweeping, changes in the way energy is bought and sold. Energy markets have undergone significant liberalization in recent years, as state energy producers have been privatized to increase competition, and electricity production has increasingly been separated, or decoupled, from delivery. As a result, since July 2007, consumers across the European Union have been free to choose between many gas and electricity suppliers. The oil industry has also been affected, undergoing dramatic changes in ownership and influence. Large integrated oil companies such as Exxon, Shell, BP, Gulf, Texaco, Mobil, and Socal (Chevron), which in the early 1960s produced, refined, and delivered most of the world's oil, have lost their control over the vast resources in the world's largest producing countries. These companies were known as the so-called seven sisters, a term originally coined by Italian oil magnate Enrico Mattei (for an in-depth history of the petroleum industry see Yergin 2008). In fact, petroleum and natural gas are now bought and sold largely through spot and futures markets.

A spot market is where a commodity is purchased for immediate delivery, and futures markets trade contracts for future deliveries. These spot and futures markets in energy commodities have become playgrounds for large investors and hedge funds, both of which have added volume and volatility to a market that is experiencing rising demand emanating from emerging economies, uncertain political risks, long-term supply pressures, and calls to reel in runaway environmental damage.

Oil in particular has moved from a risk industry to a speculators' market. Some investors use oil futures in much the same way they use gold, as a mechanism to hedge swings in the currency markets. Of course, since hedge funds remain unregulated, the actual amount of money invested in energy commodity futures is unknown, which in itself adds another variable to the uncertainty surrounding price stability. If, as a result of any or all of these factors – whether supply constraints, hedged investments, natural disasters, or changes in public policy – the price of energy rises, household disposable incomes decline, consumers grow weary, and elected governments grow nervous. Indeed, a downturn in disposable income does not bode well for politicians who need to be re-elected.

Despite all the uncertainties surrounding energy and the future, one thing remains evident. The basics of energy economics are not likely to change anytime soon. Still, the energy mix and the rules of how energy is traded just might. Europeans of the future will still drive cars, albeit more efficient ones, and in all likelihood those cars will run on biofuels, pure electricity, hydrogen, or a hybrid of them all. Parisians will still rely heavily on subways, Romans on scooters, and everyone on airplanes. Yet the more expensive and, thus, the less efficient certain energy services become, the likelihood that consumers will demand alternatives will grow; and while economics may force change, Europe's politicians are likely to determine the speed at which that change occurs.

The three end uses of energy

Decoupling energy policy from swings in commodity prices, and focusing on strategic, stable supplies of energy services is one of the principal energy challenges for public policy in the twenty-first century. To grasp this challenge one needs to look carefully at the markets and end uses for energy services. Therefore, we turn to the three main energy services, electricity, heating, and transportation, and how each functions. We begin with electricity because its reliable availability and ubiquitous use is essential to any modern economy. We follow with an examination of heating. Heating services are so vital to the welfare of the population and the functioning of industry that no energy policy would be complete without it. It also has particular importance for the diversification and security of supply because heating by electricity tends to be very expensive, demand notably spikes in winter, and it is largely dependent on fossil fuels such as coal, oil, and natural gas. We then turn to transportation, where the production and delivery of the fuel necessary to power vehicles tends to be one of the most controversial of all energy topics. One cannot ignore the sheer value of mobility to the economy or the political and economic hazards associated with its primary fuel, oil. Each section explains the mix of fuels, or lack thereof, required by each energy service, addressing supply, production, and consumption patterns, the prices paid by EU consumers for those services, and the infrastructure in place to deliver them.

Electricity

Electricity is the most commonly used and yet most difficult energy topic to understand. It is safe to say that the average European takes electricity for granted. Lights are turned on and off at will; a dead refrigerator is a small disaster, but can be quickly replaced. Virtually every aspect of a modern economy, from the subway in Paris to the lights along Germany's autobahn is electricity-dependent, and in its absence, the world as we know it would quickly be thrown back centuries. Energy analysts need to comprehend the domain of electricity because it plays such a pervasive role in modern economies. They need to know not only the source fuels used to generate electricity but also the pricing standards and differences in consumption patterns among populations and industries, and across different countries.

To begin with, electricity is not a prime energy source. Instead, it is converted energy from a prime energy resource. Electricity and electric energy are two distinct terms. Electricity is measured in units of power, and power is the rate at which electricity is generated or consumed. It is usually measured in watts, but joules are increasingly used for reporting purposes. If one million homes each turn on one 40-watt light bulb, they are consuming 40 million watts (or 40 megawatts). The terms *load* and *demand* refer to the maximum capacity of an electric power plant or network of interconnected plants, connected over what is called a *grid*.

By contrast, electric energy refers to the amount of electricity consumed over time by an end user, calculated and charged in watt-hours. Thus, demand is always listed in watts (or kilo-, mega-, giga-, or terawatts, depending on their volume) and energy is always listed as a product of use over time, namely some factor of watt-hours. Electricity bills list energy consumed, and do so in kilowatt-hours, listing a unit price for each hour consumed. Modern households loaded with commercial electronics consume on average between 2,000 and 15,000 kWh of electricity per year. As indicated by the relatively large range, there are huge differences in electricity consumption between countries, which are the result of a number of factors ranging from economic standards to cultural habits. In the United States for example, where households tend to be spacious and building materials are less energy-efficient than in Europe, residents rely heavily on air conditioning in the summer and heating in the winter, raising the average annual electricity consumption by an American household to over 10,000 kWh. By contrast, in Europe, where air conditioning is rarer and heating tends to be focused on spaces occupied at specific times, the average household electricity consumption tends to be in the range of 3,500 to 7,500 kWh per year.

In addition to distinguishing between energy consumption and electricity consumption, one should be aware that electricity-related standards vary between countries. Travellers inside the EU will attest that different member states use different standards for plugs and consumer devices. Italian, British, and Danish plugs are three-pin variants, while most of the rest of Europe use a two-pin configuration. European consumers receive electricity in their outlets at an electric potential of 240 volts. North American and Japanese consumers receive theirs at a potential of 110 volts.

Electric power plants generate electricity at varying potentials, usually different than that consumed. Once generated, electricity is transmitted along lines to transformers that increase or reduce voltage and distribute the final potential to homes and industrial centres. Electricity can be delivered in direct current (DC), as from batteries or used in electronics, or alternating (AC) current. Alternating current is usually generated at higher voltages than direct current, and thus is better suited for long-distance transmission and commercial use. Finally, if all this were not enough, there are joules (J), which is the international standard for measuring heat, electricity, and mechanical work, and British thermal units (Btu), which is the UK standard for measuring the amount of heat required to raise the temperature of one pound of water by one degree Fahrenheit.

Recognizing the terminology is important for two reasons. First, because understanding energy terms and the relationships between them helps us to compare the cost of different energy sources, whether they are prime resources, such as oil and gas, or converted ones such as electricity, and second, because it helps us to identify cost-effective alternatives.

For example, one can compare coal and oil with electricity by using approximates, with the caveat that variations in the technology used can increase or decrease conversion efficiencies. First, one needs to identify their energy equivalents, and then compare their price efficiency. Using British thermal units, joules, and kilowatt-hours, one can establish the energy equivalencies of different energy sources. One kilowatt-hour of generated electricity is roughly equivalent to 3,412 Btu or 3.6 megajoules (MJ), while one metric ton (MT) of steam coal can produce circa 22,877,388 Btu or 24,137 MJ or 6,705 kWh, and one barrel of oil can produce circa 5,800,000 Btu or 6,119 MJ or 1,700 kWh (c. 6,889 by Platts; see Hofstrand 2008, Platts 2014). Comparing these energy resources in a volume-based, petroleum-centric fashion, known as barrels of oil equivalent (boe), we can see that one metric ton of coal contains about the same energy as almost four barrels of crude oil. It is important to note that there are differences in actual primary energy content and end-use equivalents based on how oil is converted into end-use energy and to what one is comparing, but one US barrel of petroleum contains roughly 159 litres. The number of barrels in a metric ton depends on the type and specific gravity of the crude oil in question. According to OPEC, one metric ton of oil is approximately 7.33 barrels; the EU27 consumed 444,570,000 tonnes of oil equivalent (toe) in 2011 according to Eurostat, which is roughly 3.3 billion barrels of oil or 12.6% of a cubic mile (for a detailed explanation of the maths involved, see Crane et al. 2010: 6–9; for CMO conversion rates, see Crane et al. 2010: xxi).

If the numbers seem daunting, imagine the following comparison that shows just how efficient oil is as an energy source. One cubic mile (4.17 cubic kilometres (CKO)) of oil (CMO), which amounts to 26.2 billion barrels of oil, produces as much energy as four of the gigantic Chinese Three Gorges Dam, 104 coal-fired power plants, 52 nuclear power plants, 32,850 wind turbines, or 91,250,000 solar power panels, each operating for 50 years (see Crane et al. 2010, Goldstein and Sweet 2007). One CMO is equal to 15,300 TWh of end-use electricity and has a primary energy equivalent of

5.13 TWy. For comparison, across all 28 EU countries, total electricity generation available for final consumption in 2012 amounted to 2,796 TWh, or about 18% of a cubic mile of oil equivalent, but its total energy consumption of over 70 million TJ in the same year equates to almost 43.5% of a cubic mile of oil equivalent.

Over the first quarter of 2014, the average price for European Brent crude oil was just over $108 (€77.78) a barrel, and a metric ton of steam coal delivered in the Netherlands was just over $78 (€56.17). Translated into energy equivalents, this means that the cost of energy from coal was roughly a sixth of that of oil, or €0.01 versus €0.06 per kWh. These prices do not include taxes, shipping, or any processing costs or service charges, but having a sense of their raw economic value is helpful, since alternative fuel sources for electricity generation can be calculated in a similar way. Learning how to make such calculations significantly reduces the complexity of identifying viable and efficient alternatives.

Across Europe, the average medium-sized household consumes between 2,500 and 5,000 kilowatt-hours of electricity per year and pays between €0.09 and €0.30 per kWh of electricity including taxes (Figure 2.1), with rates varying according to overall annual use. (Eurostat provides data on electricity prices for consumers on its website.) This means that the average medium-sized European household pays somewhere between €231 and €1,500 for electricity every year. This gap in consumer electricity prices is just one of the many energy-related issues (internal dimension) under increasing focus from EU efforts to integrate their energy policies.

Once produced, electricity has to be transmitted from a power plant to a grid, where it is either immediately consumed by end users or sent along high-voltage lines to another grid covering another large area. If something goes wrong, such as a wire severs or a power plant fails, the loss of load has to be immediately matched by another power plant on the grid. Failing that, consumers suffer a reduction in capacity, known as a brownout, or a complete elimination of electric power, known as a blackout. Such events are not rare. In September 2003, Sweden and Denmark experienced massive losses when a 1.1 GW nuclear reactor stopped functioning, causing another 1.8 GW reactor to do the same, leaving five million people without electricity. Only days later, Italy experienced a similar incident. The events of 2003 were largely the result of flaws in the physical energy infrastructures, common in large integrated economic zones (see Bialek 2004).

Figure 2.1 *Consumer electricity rates (€/kWh, 2013, 1st semester). Medium-sized households, consumption = 2,500–5,000 kWh/annum*

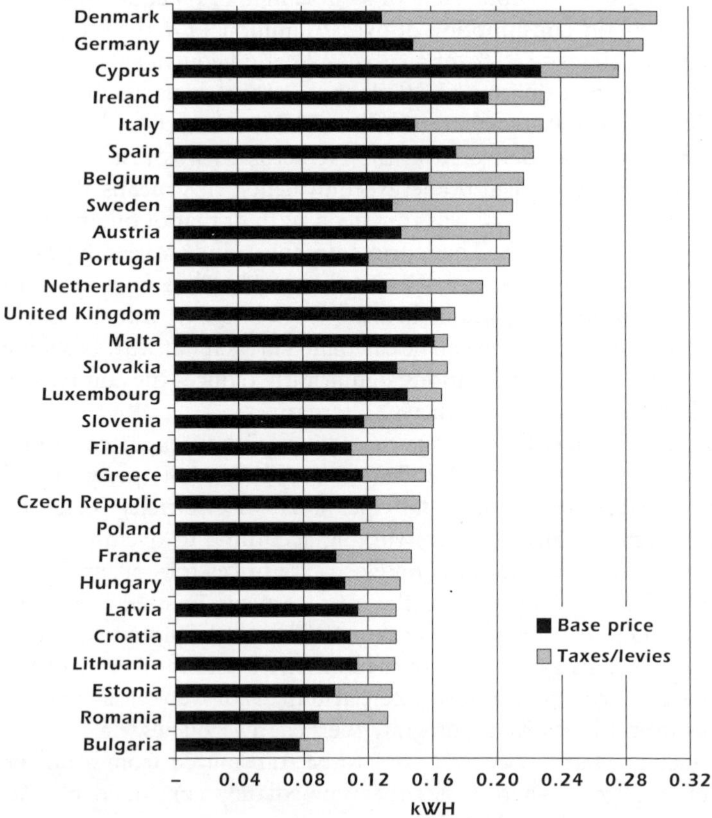

Source: Based on Eurostat (2014q) data

Importantly, the EU does not share one common electricity grid. In fact, the member states operate independent grids, developed at different times, using different voltage levels, and a mix of direct and alternating currents. Europe's electrical network system is complex, but its logic is simple. Electric power plants feed the grid with electricity according to current demand. Since demand fluctuates daily across different grids in different countries, excess electricity supplies are exported and shortfalls are met by imports. As a result, the trade in electricity within the EU is huge. In 2012, for example, six of the world's top ten exporters of electricity were member states of the EU, and Germany exported 66.6 TWh of electricity to the European network (importing 43.8 TWh) (Statistisches Bundesamt 2013).

As of 2014, the European electrical network substantially integrated, representing a good example of how the EU has successfully converted some energy issues from external to internal dimensions. Divided regionally, the network is coordinated by the European Network of Transmission System Operators for Electricity (ENTSO-E), which was initiated with the adoption of the EU's Third Energy Package. Europeans have been working extensively to integrate their electricity markets and upgrade their grid systems to allow small, local producers to add alternative energy-sourced generation to the overall market, so reducing transmission distances and increasing efficiency by localizing production and consumption. These smaller integrated networks, known as Smart Grids, are a budding feature of modern electricity networks.

Electricity produced in one country is frequently exported to another, irrespective of how it was generated. When a light is turned on in Munich, it is virtually impossible for the consumer to control whether that electricity came from an environmentally friendly renewable source, such as hydropower, or an environmentally hazardous one, such as the burning of coal. Reducing fossil fuel consumption in the electricity sector, therefore, requires a reduction in use of fossil fuels as feedstocks in electricity generation.

The largest single consumer of electricity in the EU is the industrial sector, which consumed 36% of all the electricity generated by EU member states in 2012. The services sector was close behind at 30.2% and households accounted for 29.6% (Eurostat 2014h). By comparison, the transport industry (including trains and subways), which relies more heavily on fuels than on electricity, accounted for less than 3% of electricity consumption EU-wide. These figures suggest that an important focus point for electricity conservation efforts should be the industrial and commercial services sectors, which together account for 66% of annual EU electricity consumption. This is not to state that household savings will have no impact, but rather that a focused effort to improve efficiency standards at the industrial and commercial level will achieve far deeper savings.

One must distinguish between electric energy use and other major forms of energy consumption. Unlike gasoline or firewood, electricity cannot yet be stored in large quantities or for long periods. Instead, it must be consumed as fast as it is generated.

Electricity produced in France may be exportable to Germany, but will experience greater losses in power the further it travels to the market, making the delivery of French electricity to Eastern Russia unprofitable and very unlikely. In order to deliver electricity from the producer to the end consumer, an extensive integrated network is required, and the amount of electricity generated is always a product of consumer demand. Overproducing electricity will not necessarily lower prices, since the extra supply will be lost as long as there is no mechanism to store it en masse.

Electricity is thus rigid in terms of the balance between supply and demand as well as geography. Finally, electricity can be generated using a myriad of sources, from renewables to fossil fuels, from the environmentally hazardous to the environmentally neutral. Therefore, while electricity is rigid in terms of supply and demand, it is also flexible in terms of how it is generated. Power sector policies thus span all three dimensions of the energy policy typology and pose particular and complex challenges to the pursuit of comprehensive energy security.

Increasingly, a new form of hybrid power plant has been coming online. These so-called cogeneration or combined heat and power plants (CHP) produce both electricity and heat, generally using a gas turbine with heat recovery, which significantly increases the efficient use of feedstock fuels such as natural gas. The use of a single plant to produce both heat and power has the benefit that consumers of electricity in any sector of the economy are usually also consumers of heat. These new plants thus reduce the need for additional fuels to be purchased by consumers for the local generation of heat, significantly reducing negative environmental impacts, such as CO_2 emissions. In 2011, approximately 11% of all EU27 electricity generation was generated using combined heat and power plants (Eurostat 2014i). As in the case of the role of renewables sources, there is wide variation in the use of CHP across the Union, with Denmark (46.2%) and Latvia (47.4%) generating almost half of their electricity this way, while Malta generates 0.9% and Cyprus none.

Heating

If one considers the wide range of choices and uses of electricity generation and distribution, then by comparison, the energy service of heating is relatively simple. Consumers rely on heating for temperature control (space-heating and water) and cooking.

Although electricity is sometimes used for the former – in 2011 electricity accounted for roughly 25% of overall residential energy consumption in the EU28 (Eurostat 2013: 62) – it tends to be used more often for cooking than for space-heating and actually accounts for a very small percentage of overall heat consumption. Instead, the two most common sources of space-heating are oil and natural gas, with wood products not far behind. Due to variations in fuel prices and availability, as well as significant differences in the average number of cold days depending on geography, the exact balance of sources used to generate heat varies significantly.

The largest demand for space-heating occurs during the cold months of winter. Most of the final natural gas consumption (45.6% in the EU is consumed by what the European Commission classifies as the 'households/services' sector (European Environment Agency 2013)) and the lion's share of natural gas use goes to heating (including hot water) and cooking (Eurostat 2007: 42–43). In Austria, the home heating market in 2010 was divided relatively equally at the national level between natural gas (26.1%), heating oil (20%), and district heating (23.1%), with a notable 16.6% using wood and only 7.3% using electricity (Österreich Institut für Wärme und Oeltechnik 2012).

Consumers who heat their homes with natural gas are charged for the volume of gas they consume based on kilowatt-hour equivalency. The average medium-sized household across the EU28 consumes a gas calorific value of between 20 and 200 GJ per year (500–2,000 m^3). Again, location makes a big difference. Including tax, a Romanian pays around €0.0285 per kWh of natural gas consumed, while a Swede pays as much as €0.1227 for the same product and use (Figure 2.2).

Unlike natural gas, heating oil provides customers with individual security, as their personal stocks are stored in household storage tanks. According to the European Heating Oil Association (Eurofuel), heating oil accounts for approximately one-fifth of total energy consumption for domestic heating and hot water purposes. As in the case of gas, prices for heating oil tend to rise gradually in the winter months when demand is highest and fall during warmer seasons. The price of heating oil is a combination of the price of oil, margin costs (refining, transportation, insurance, stockpiling, distribution, and sales to consumers), and taxes. The distribution of these costs varies by country. In the US where taxes are relatively low, the crude oil itself accounted for 65% of the price in 2011 (EIA

Figure 2.2 *Household consumer prices for natural gas (kWh, 2013, 1st semester). Medium-sized households (Band D2), consumption = 20–200 GJ/annum*

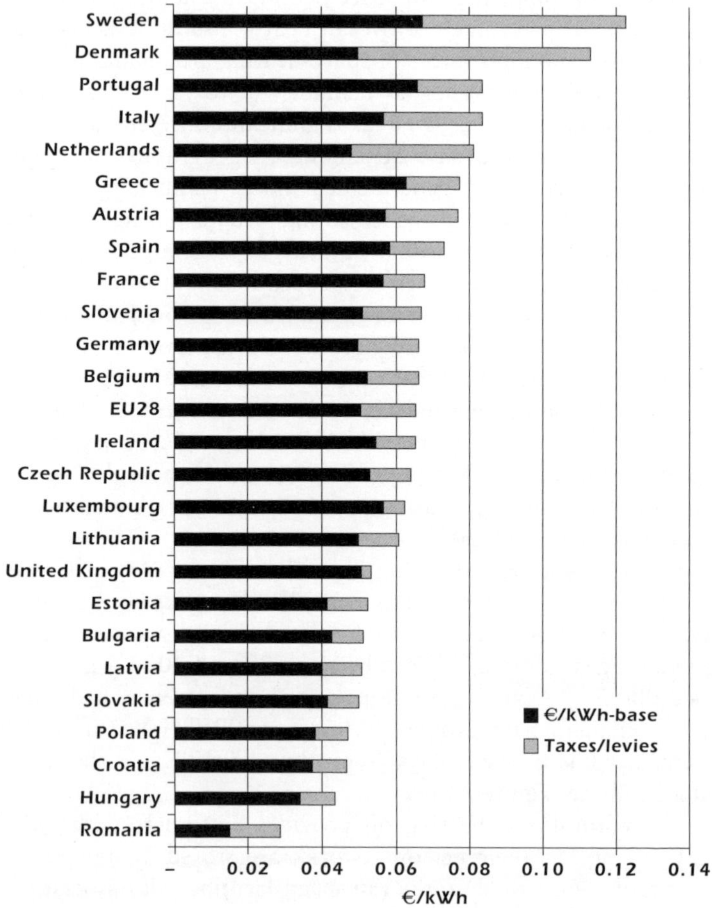

Source: Based on Eurostat (2014r) data

2013). The average distribution of costs in the retail price of heating oil in Europe in May 2014 was 47% for the crude oil, 21% for the margin, and 33% in taxes. Since heating oil is a refined petroleum product, the cost of heating with oil rises and falls with the price of petroleum. Again, geography matters. Across the European Union, one litre of heating oil ranges between €0.595 and €0.905 before taxes, and between €0.768 and €1.565 when all taxes and duties are included. Citizens of Luxembourg pay just under half per litre, as do citizens of Denmark (see Figure 2.3).

In summation, heating is a service largely dependent on fossil fuels such as coal, oil, and natural gas, one in which demand notably spikes in winter, but also one where consumers can choose between different fuels and delivery methods. As the EU has to import large amounts of these resources, particularly in winter, and burning them is detrimental to the environment, policies that affect pricing

Figure 2.3 *Home heating oil prices (€/litre, as of 6 May 2014)*

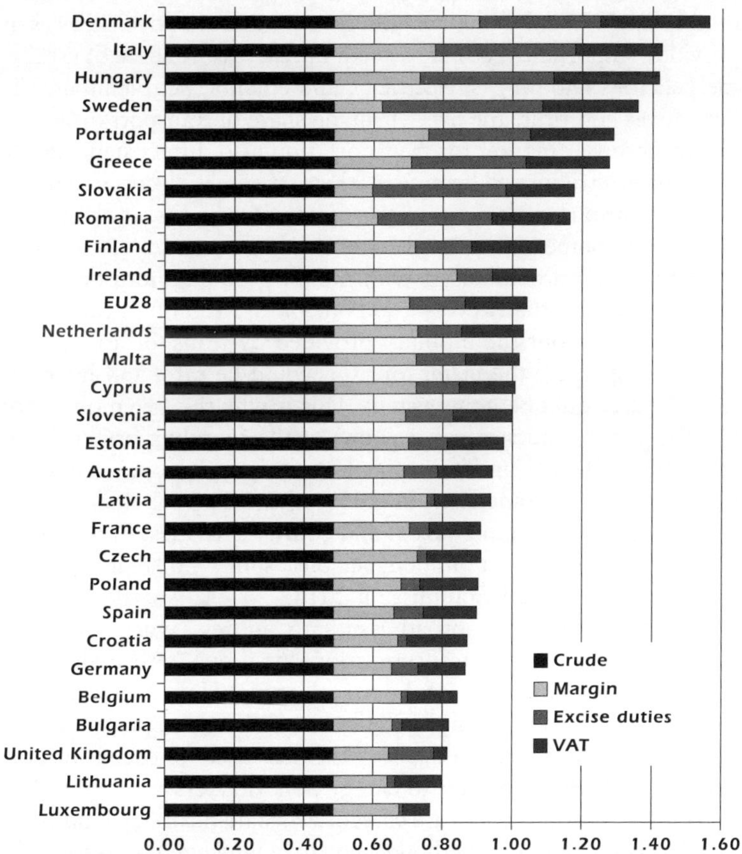

Source: Based on Europe's Energy Portal (2014) data

in the heating sector cover both the external and multidimensional domains. Indeed, heating Europe's homes and businesses presents serious and recurring challenges to policymakers trying to guarantee security of supply, fair competition, and environmental sustainability – the three pillars of comprehensive energy security.

Transportation

Transportation by itself is not a final energy service, but rather a sector dependent upon the price and availability of energy supplies to power vehicles. While similar statements could be made of electricity (electric energy is used to power devices) and heating (fuels are burned to produce heat), the transport sector is hampered by a lack of substitutable fuels and by the magnitude of the derivative services it provides. Importantly, despite a small but increasing share of useable fuels such as natural gas, electricity, and ethanol, petroleum and its derivatives constitute the overwhelming share of transport fuels.

The end-user market for transportation also differs significantly from the electricity and heating markets. Whereas electricity largely powers stationary devices and heating is used for cooking and closed-environmental control, and both are end-user services, transport fuels provide users with mobility, adding both flexibility and costs into virtually every good one buys. Mobility is the operative term. Without the mobility provided by transportation, food would never leave the farm, tourism would be rare, and international trade would be non-existent. This is why the free mobility of people, goods, and services was one of the founding principles of the Treaty of Rome, and it is difficult to envisage Europe's economic success without transport. In 2010, the transport and storage services sector (including postal and courier activities) in the EU27 employed around 11.1 million persons, some 5.0% of the total workforce (European Commission 2013: 19).

There are thus two broad issues to address when analysing energy services used for transportation. First, one must understand the scope and complexity of the overall sector. How important is it to a well-functioning economy? Who or what is transported, where and how? Second, one must specify the fuels used for each transport activity, and why it matters to energy policymakers.

Transportation services are categorized statistically by load (who or what is moved) and method (the way in which a load is moved), sometimes referred to as mode. There are two load categories, passenger (people) and freight (goods), and five methods (modes), including rail (train), road (cars, motorcycles, buses, and trucks), inland waterways (riverboats), maritime (seafaring vessels), and airplanes. A sixth method, pipeline, is also used to transport fluids and gases, predominantly fuels, but also water. Importantly, and with the exceptions of pipelines and electric-powered passenger rail, refined petroleum products fuel virtually every mode of transport.

One calculates macro figures for passenger and freight transport in kilometres travelled. For example, in 2011, European passengers used their cars to travel over 4.8 thousand billion passenger-kilometres (pkm) in total. By comparison, Americans in 2010 drove cars some 5.9 thousand billion pkm. Most passenger transport in the EU and the US occurs by passenger car, with airplanes accounting for the next largest share. The overall use of subways, trams, and metros is considerably lower by comparison. The figures and preponderance of road usage for freight is also notable. Both the EU and the US used roads in 2011 to transport goods over 1.7 thousand billion tonne-kilometres (tkm) (European Commission 2013: 36). In 2006, seafaring vessels travelled another 1.4 thousand billion tkm to deliver goods from one EU port to another. Still, in Europe and the US, passenger transport outdistanced freight by a factor of no less than 1.5, and passenger car travel outdistanced freight moved by all modes taken together over the same period.

According to Eurostat, intra-EU and domestic land transport averaged just over 13,060 km per person in 2011. Passenger cars accounted for more than 73.4% of that travel (ibid: 19). Trains, trams, and subways accounted for only 7.7%. Even if one adds into the equation intra-EU air transport, which amounted to 8.8% in 2011, as well as another 0.6% for passenger sea travel, reducing the share of passenger cars, the significance of the passenger car cannot be overstated. By an overwhelming proportion, cars provide the primary means of human transport, whether that travel is for recreational or economic purposes (Figure 2.4) (see also Eurostat 2014j).

The fact that passengers, and specifically passenger cars, account for the largest share of transport activity, as seen in kilometres driven, is also reflected in another two categories of comparison: vehicles and infrastructure. In 2010, the EU had five million kilometres of paved roads, and in 2011, a stock of 242 million passenger cars and 34 million commercial freight vehicles. This compares with the US, which in 2010 had 4.4 million kilometres of paved roads, a stock of 230 million passenger cars, and almost 11 million

Figure 2.4 *Passenger transport by mode (2011)*

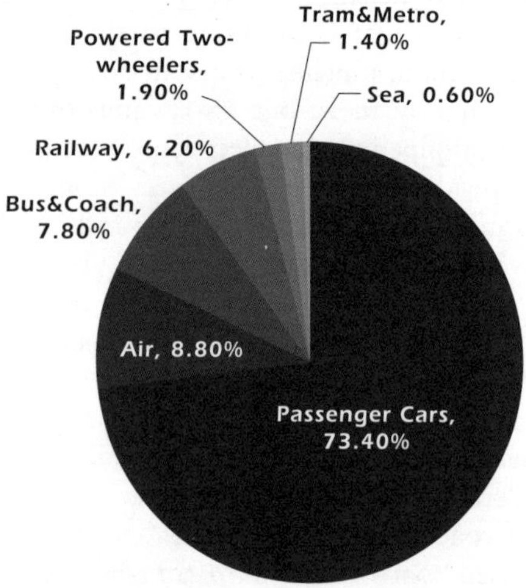

Source: European Commission (2013: 48)
Note: Minor discrepancies are due to rounding issues.

commercial freight vehicles. By comparison, Russia, one of Europe's primary suppliers of petroleum, had only 786 thousand kilometres of paved roads in 2010, and fewer than 35 million cars, and 5.5 commercial freight vehicles in 2011 (ibid: 33).

There are of course variations in the use of cars versus other means to travel, depending on geography and national infrastructure. Hungarians use cars less often (63.4%) than Lithuanians (90.8%), and the French choose to use rail services (10.3%) about as often as Austrians choose to use buses and coaches (10.1%). Still, across the EU27, the tendency to choose cars over other means of personal transport in 2011 was strong and did not dip below 63%.

The use of road vehicles is also the preferred choice for transporting goods within Europe. In 2011, Europe's roads were used to carry 45.3% of all intra-EU27 freight and 71.8% of all inland freight (from port to end point, Figure 2.5). Freight trains carried only 11% of transported goods in the same year. By contrast, in the US in 2009, trains accounted for 42.2% of all inland freight transport, while roads carried only 35.2% (European Union 2013: 44).

In terms of Euro value, EU external commercial trade relies more on sea than roads (Figure 2.6). However, taken together sea and road transport accounted for a total of 68.5% of imports and 68.1% of exports. Pipelines, by contrast, only account for 6.2% of

Figure 2.5 *Modal split of freight transport in the EU (%, 2011)*

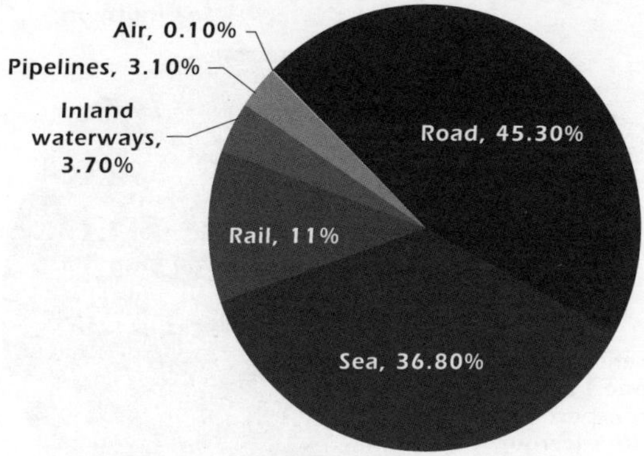

Source: European Commission (2013: 36)

Figure 2.6 *EU imports/exports by mode of transport (2011)*

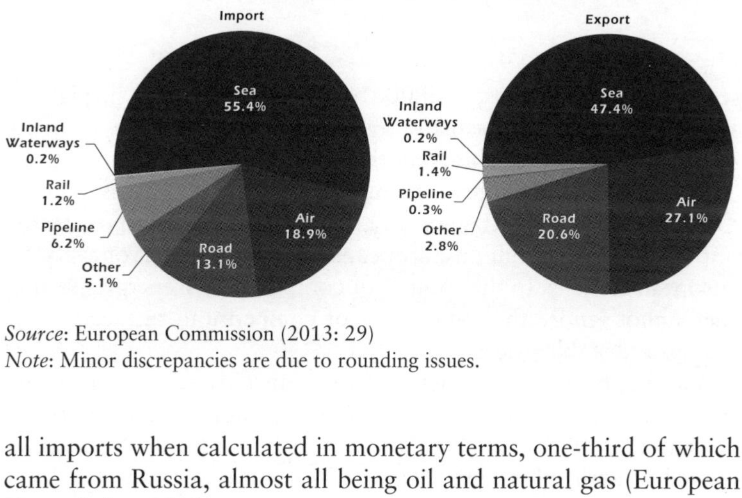

Source: European Commission (2013: 29)
Note: Minor discrepancies are due to rounding issues.

all imports when calculated in monetary terms, one-third of which came from Russia, almost all being oil and natural gas (European Commission 2013).

As illustrated above, Europe is dependent on transport for everything from trade to recreation. It can easily be seen how policies applied to this sector span all three energy dimensions. However, transport is not merely a method of mobility. There are more than one million companies operating in the transport sector, and those working in road transport accounted for 33% of the sector's annual turnover (European Commission 2013). While means of transport such as airplanes and sea vessels employed significant numbers of

Figure 2.7 *EU27 transport sector employment by mode, 2010*

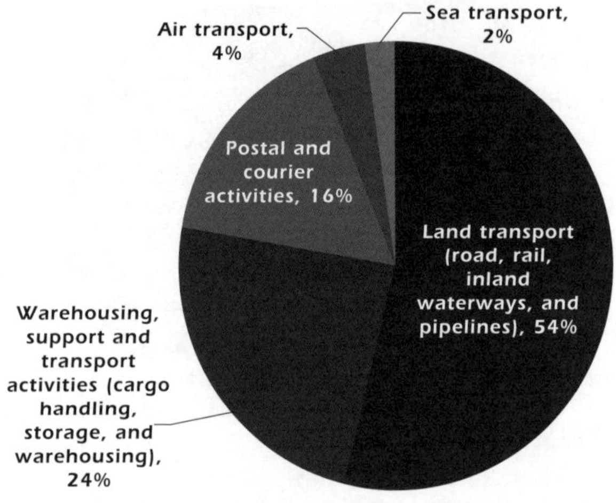

Source: European Commission (2013)

people in 2010 (almost 600,000 worked on ships or airplanes), nearly 5 million Europeans derived their gainful employment by driving trucks and cars, and another 2.5 million people worked for the sector's second largest operation, auxiliary functions such as handling and storage (Figure 2.7). Altogether, the transport sector employed over ten million Europeans in 2010, or 1 out of every 50 European citizens. In the context of comprehensive energy security, one cannot ignore the importance of competition, the security of supply, and stable prices.

Also important is the fact that the transport sector is growing and, indexed on 1995 figures, has been doing so consistently through to 2008. While the economic crisis that hit the EU in 2008 slowed growth, passenger transport (in pkm) still grew by 0.8% between 2000 and 2011 and freight transport (in tkm) grew by as much as 0.9%. While such statistics demonstrate transportation's importance for delivering goods and passengers in terms of mobility, it also demonstrates the importance of such activities to Union-wide employment and a thriving economy.

The number of people travelling by air, while substantially less than by road, has also experienced explosive growth in recent decades. According to Eurostat, roughly 200 million passengers travelled to, from, and between the EU15 by air in the mid-70s. In 2012, over 827 million passengers flew to or from the larger EU27 group, including interregional travel, and approximately half of those passengers (350 million) flew to and from destinations within the Union, with Amsterdam/Schiphol being the busiest airport for intra-EU traffic (Eurostat 2014k). Between 2011 and 2012, the number of internal EU27 air passengers rose by 0.7% and seen in terms of distance travelled, the demand for intra-EU passenger air travel in 2011 is estimated to have been around 578 billion pkm; and in 2012, airplanes moved over 2.2 million tonnes of freight and mail within the Union (European Commission 2013).

Since the amount of fuel used per flight is a factor of weight, distance, and the efficiency of the aircraft involved, it is very difficult to estimate the total energy consumption of the air transport sector or its overall efficiency. However, Eurostat reported that of the 364.08 million tonnes of oil equivalent (Mtoe) consumed by the transport sector in 2011, air transport accounted for 14% or 50.5 Mtoe. This is particularly interesting since air transport only accounted for 8.8% of passenger travel, less than 1% for tonne-kilometres of freight transported, and one-eighth the total distance travelled by passenger cars (575 billion pkm vs. 4,822 billion pkm) over the same period (European Commission 2013).

According to the International Air Transport Association (IATA), modern commercial aircraft achieve fuel efficiencies of around 3.5 litres per 100 pkm. Newer aircraft like the Airbus 380 and Boeing 787 achieve closer to 3 litres per 100 pkm, which IATA (2014) claims is 'better than a compact car'. Of course, commercial aircrafts never carry one passenger, but given the distance they travel and the savings in time and cost-effectiveness for the trip, air travel is one of the more efficient, if not environmentally destructive, means by which to travel.

The cost of transportation is an important factor in the energy question because it is directly related to the cost of fuel, which is paid out of consumer pockets, reducing available disposable income for other goods and services. In fact, transportation accounts for a significant share of household expenses. In 2012, European households spent roughly €967 billion of their total consumption on transport, or between 7.7% and 19.1% of their annual budget including new purchases, operations (including fuel purchases), and servicing (Figure 2.8). Just as in the case of electricity and heating, budgetary allocations vary depending on geography, with the

Figure 2.8 *Final consumption (as %) of household expenditure (2012)*

Source: Based on Eurostat (2014l) data

Slovaks spending 7.3% and Lithuanians spending just over 19%. Still, regardless of the country in question, passenger cars constitute the bulk of household transportation costs.

Since the use of road vehicles is so prevalent, and since transport is such a significant share of household consumption, employment, and trade, energy policymakers are correct to be deeply concerned with issues of supply and price stability of the fuels used to operate those vehicles.

As is the case with heating oil, gas, and electricity, gasoline for automobiles and diesel for cars and trucks vary significantly depending on where the fuel is sold (Figure 2.9), and taxes play an important role (and provide substantial income for member-state governments). Polish consumers for example paid a final tax-inclusive price as little as €1.27 for a litre of Euro-super 95 in April 2014, while the Dutch, Danes, and Italians paid between €1.71 and €1.76. For diesel, the Luxembourgers paid the least at €1.18 and the British, who produce a lot of oil, paid €1.64. Fuel taxes also vary according to the country. For example, in April 2014, 48% of the retail price that Luxembourgers paid for Euro-super 95 is tax (and 42% for diesel), while Dutch consumers paid 60% and 50%, respectively, for the same fuels. In fact, Europeans were paying taxes of between 82% and 175% on top of the base fuel price for every litre of unleaded gasoline in early 2014, and 71% and 147%, respectively, for diesel.

Finally, just as the fuels used to power cars and trucks are derived overwhelmingly from petroleum, notwithstanding the increasing amount of biofuels that make up a litre of gasoline or diesel. Commercial sea vessels rely on heavy fuel oil and airplanes on high-octane kerosene, neither of which as of yet has a commercially viable substitute. Therefore, substitution in the transport sector is one of the biggest energy challenges faced by policymakers. Whereas heating and electricity are both energy services that can be derived from multiple sources, the same cannot be said of the energy services needed for mobility. The EU27 consumed 444.57 million tonnes of oil in 2011, of which almost three-quarters (364.1 Mtoe) were consumed in the transport sector (European Commission 2013). That amounts to more than 2.6 billion barrels of oil every year, and, in 2012, the transport sector accounted for approximately 25% of all greenhouse gas emissions emanating from the EU at circa 1,039 million tonnes of CO_2-equivalent (European Environment Agency 2013a).

New technologies are under development. Hydrogen, electricity, biofuels, and hybrids integrating any of the three are slowly entering the end-user transportation market. However, the replacement of

Figure 2.9 *Per litre prices of Euro-super 95 and diesel, 13 April 2014*

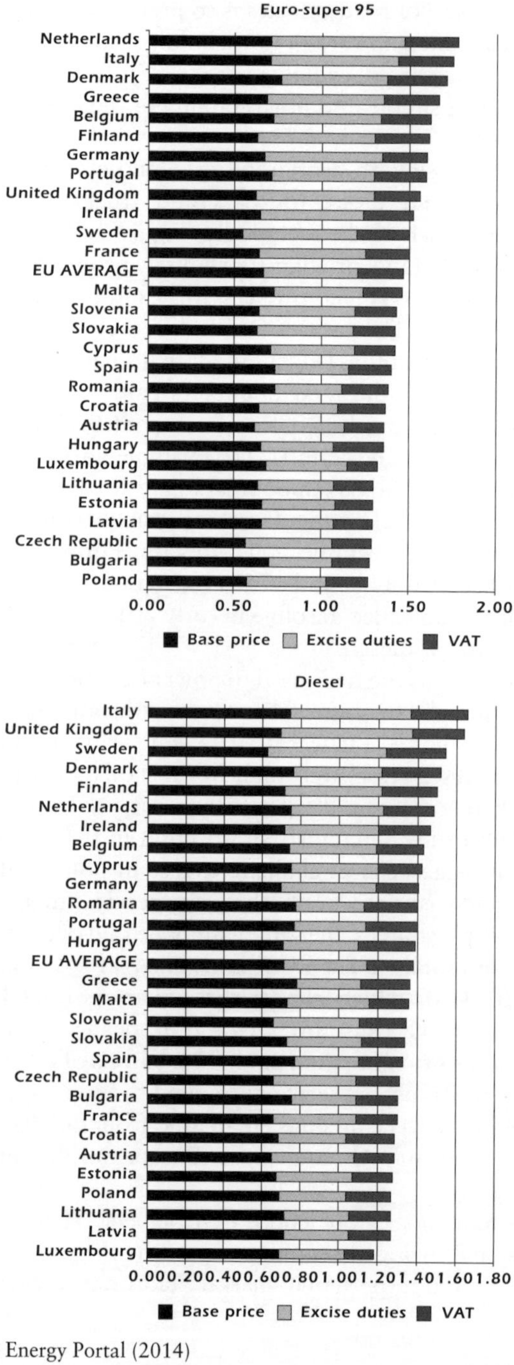

Source: Europe's Energy Portal (2014)

petroleum fuel by an alternative power source for the hundreds of millions of vehicles currently in use will require a massive overhaul of the transportation energy infrastructure, if not a legislative revolution.

Therefore, for the time being, oil remains the primary source of fuel for individual mobility and economic activity. As a result, the reliable availability of oil (i.e. the security of its supply and stability of its price) is an essential factor in maintaining economic stability. The best way to guarantee supply security and price stability is to own the resource; the only alternative is to acquire it through trade. Europe, however, owns precious little, forcing it to import vast amounts of coal, oil, and natural gas, raising the geography of energy resources (the external dimension) to a matter of utmost importance.

Global balance of supply and demand

The three main energy services, electricity, heating, and transportation, require resources that are not evenly distributed between originating countries. The following section looks in depth at Europe's available fuel sources, identifying its most important suppliers and trade patterns, and provides information that is vital to understanding the geography of energy as an aspect of international trade. It provides a picture of which fuels are consumed by which sectors and from where they come, revealing degrees of foreign dependence. Because energy is traded globally and there is fierce competition for energy resources, Europe finds itself in competition with every other major consumer on earth. Comprehending the geography of energy is a prerequisite to assessing the risks and rewards of different energy policy options.

To begin with, as Figure 2.10 demonstrates, the EU was the world's third largest consumer of energy in 2012 at 13.4% of all global consumption, just behind China and the US at 21.9% and 17.7%, respectively (BP 2013). The consumption patterns of the EU show its enormous dependence on fossil fuels (Figure 2.11) and, as Figure 2.12 clearly shows, transportation is the dominant sector next to industry in terms of sectoral consumption.

In 2012, the EU28 depended on foreign sources for 53.3% of its overall fuel needs. It was 42.2% dependent for solid fuels such as coal, 86.4% for oil and its refined derivatives, and 65.8% for natural gas (Eurostat 2015). In the case of oil, the EU27 imported some 6.6 billion barrels in 2011 (European Commission 2013b) and circa 6.5 billion barrels in 2012 (Eurostat 2012a). Figure 2.13 shows that almost half (48.5%) of that oil in 2011 came from countries where

Figure 2.10 *Final world energy consumption (2012)*

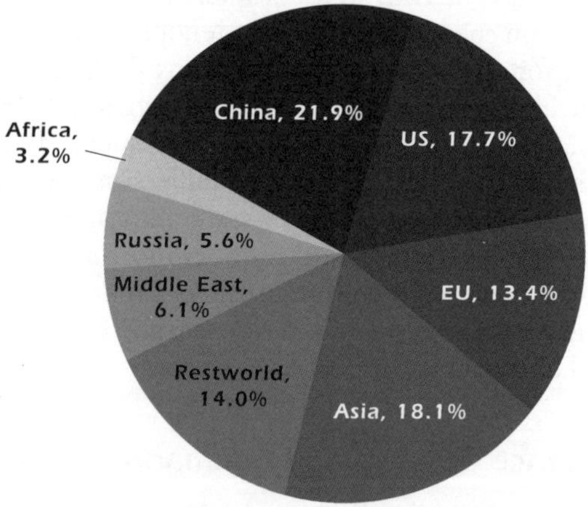

Source: Based on BP (2013) data

Figure 2.11 *Final energy consumption (TJ) for EU28 by fuel, 2012*

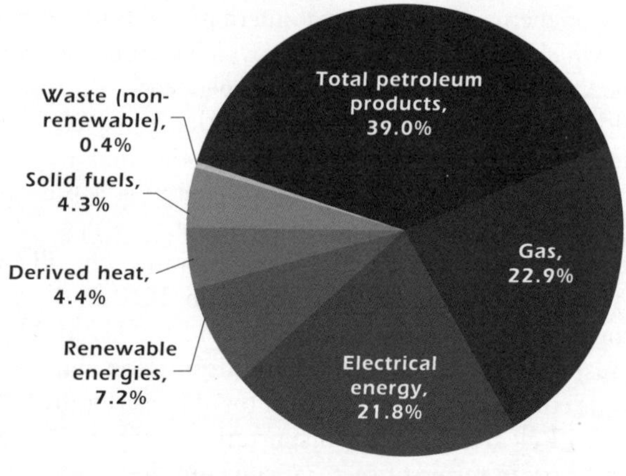

Source: Based on Eurostat (2014m) data

the record of economic diversity and democratic freedoms are less than stellar (34.8% from Russia, 8% from Saudi Arabia, and 5.7% from Iran). Meanwhile, Russia and South Africa supplied the EU with half of all its coal imports in 2011, and the United States and Australia accounted for an additional 25%. It is notable that the amount of coal

Figure 2.12 *Final energy consumption (Mtoe) for EU28 by sector, 2012*

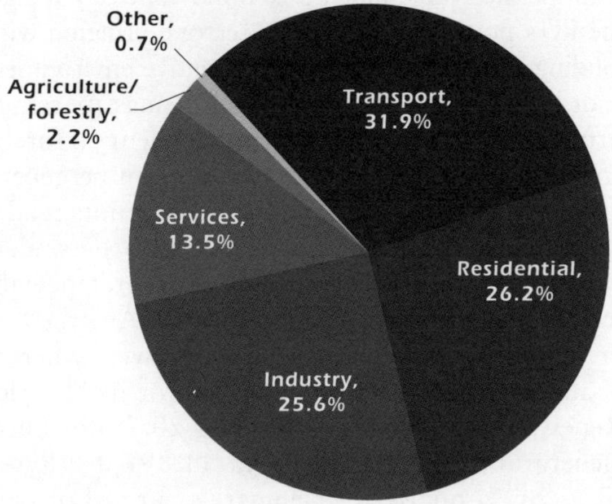

Source: Based on Eurostat (2014n) data
Note: Minor discrepancies are due to rounding issues.

Figure 2.13 *Crude oil and natural gas imports by origin, 2011*

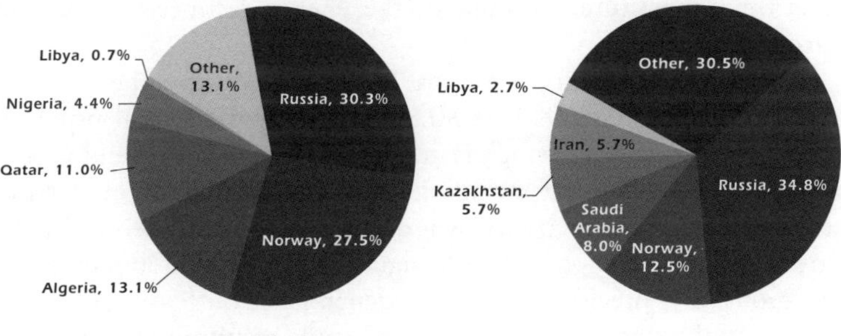

Source: Eurostat (2013: 44)
Note: Minor discrepancies are due to rounding issues.

supplied annually by Russia has almost tripled since 2000 (Eurostat 2013) and a glut of cheap natural gas in the US (the result of its shale gas boom) has freed up vast volumes of cheap American coal for markets in Europe. As a result, the US became the EU's second largest supplier after Russia in 2012, a trend that seems likely to continue for some years. The seemingly fixed nature of the EU's lopsided dependence on external supplies for fossil fuels reveals the importance of the external dimension of EU energy policy, but so too does its evolving roster of suppliers. Indeed, based both on market and political forces, it is not all inconceivable that US coal may one day replace Russian gas in the EU's power and electricity sectors, bringing with it all the corresponding economic benefits and negative environmental consequences; or better said, policy decisions with massive ramifications.

Importantly, the EU is slightly less dependent on foreign sources for electricity generation. Since electricity can be generated from multiple sources, and because physical constraints tend to compel a maximization of local generation, foreign dependence varies greatly according to the local availability of renewable energy resources. For example, countries such as Austria and Sweden rely heavily on locally available hydropower, whereas Cyprus depends almost entirely on imported oil for its electricity generation. Renewable sources accounted for 20.5% of Europe's electricity generation in 2011, up from 11.8% in 1990 (Eurostat 2013). Roughly, a quarter of Denmark's, a third of Latvia's, and over half of Austria's gross electricity consumption comes from renewable sources. Nevertheless, when taken as a block, in 2012, 52% of Europe's electricity continued to be generated from conventional thermal sources (primarily coal and natural gas). Only France relies largely on nuclear materials. Depletable fossil fuels thus constitute the bulk of the energy resources in use in Europe.

Similar conditions apply to most of the world's developed economies. Figure 2.14 shows that 80.9% of the world's total primary energy supply (TPES), which is 'all energy consumed by end users, excluding electricity but including the energy consumed at electric utilities to generate electricity' was derived from fossil fuels in 2011. Down slightly from the 86.6% figure of 1973, but nonetheless enormous. It is precisely this global dependence on fossil fuels, and the essential value of the international trade resulting from it, that restricts the ability of policymakers to define and implement working energy policies. A major shift in consumption patterns away from fossil fuels directly affects trade, and likewise, a breakdown in the international trade of fossil fuels could lead to domestic economic disaster. Therefore, understanding the trade in energy resources, specifically who produces and who consumes, is essential to any analysis of energy politics.

Before investigating which countries or regions are in control of the world's energy resources, it may be useful to get an initial overview of

Figure 2.14 *World total primary energy supply by fuel (Mtoe), 1973/2011 (excluding electricity and heat trade)*

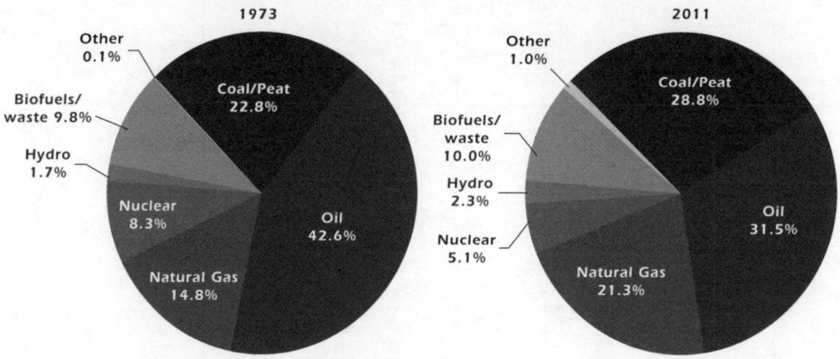

Source: International Energy Agency (2013a: 28)
Note: Minor discrepancies are due to rounding issues.

the world energy balance. Table 2.1 provides a breakdown of how different fuels balanced out on a global scale in 2011.

The world energy balance in 2011 reveals a major energy dilemma: given the enormous production and consumption of finite resources (only 13.27% of total TPES and 33.64% of TFC were derived from renewables). How long will such limited resources last?

According to BP's 2013 *Statistical Review of World Energy*, assuming no new resources are discovered and production and consumption levels stay constant (the so-called reserves-to-production ratio, R/P), the world has 52.9 years of petroleum, 55.7 years of natural gas, and 109 years of coal remaining.

New discoveries are bound to be made, and both supply and demand fluctuate, limiting the significance of such data. Projected oil supplies have hovered in the 40–42 year range for the last two decades (Table 2.2). However, new discoveries are becoming more expensive and complicated with time. Companies increasingly have to search in deeper waters and more remote locations. Drilling platforms are expensive, and the facilities and agents to build them are limited. Together with prevailing uncertainty over economic stability, decline in demand, and legislated shifts to alternative fuels, the financial risks of exploration are increasing. Unless demand declines significantly, something considered very unlikely without the massive introduction of substitutes or a global economic collapse, new discoveries will cost more, rendering stable R/P ratios as a possible indicator of long-term increases in primary energy

Table 2.1 *World energy balance in Mtoe (2011)*

	Coal/peat	*Crude oil*	*Petroleum products*	*Gas*	*Nuclear*	*Hydro*	*Combust. renew./waste*	*Other*[a]	*Total*
TPES	3,776.06	4,219.57	–83.58	2,786.95	674.01	300.17	1,312.15	128.05	13,113.38
TFC	903.62	18.76	3,614.51	1,380.50			1,111.74	1,888.41	8,917.53
which divided into the following sectors:									
Industry	728.93	10.67	312.48	506.38			198.15	800.14	2,556.74
Transport[c]	3.41	0.02	2,265.21	92.52			58.61	25.16	2,444.94
Other	132.02	0.50	435.55	610.23			854.99	1,063.11	3,096.43
Non-energy use	39.22	7.56	601.27	171.36			–	–	819.42

Total Primary Energy Supply (TPES) = production + imports – exports (+/–) stock changes
Total Final Energy Consumption (TFC) = TPES – (sum of uses listed above). TFC should balance with usage listed below.

[a]Other includes geothermal, solar, electricity and heat, wind, etc.
[b]Also includes patent fuel and BKB plants
[c]Includes international aviation and marine bunkers

Source: International Energy Agency (2013a: 37)

prices, and thus-end user services. Using the same method, a similar analysis might be made for the two other primary fossil fuel resources of coal and natural gas (Table 2.2).

Debate over the long-term supply of the world's fossil fuels is exacerbated by their uneven geographic distribution. If controlling reserves is important, then producing and then exporting them is even more so, and the latter two are not necessarily synonymous. For example, according to BP, in 2012, the United States was the world's third largest producer of petroleum in millions of tonnes (Mt) at 9.6%, behind Russia (12.8%) and Saudi Arabia (13.3%). However, the United States was the world's largest importer of oil in 2012 (19.1% of all oil imports), importing as much as Japan and China combined, indicating the country's enormous level of consumption. Europe accounted for over 22.6% of global oil imports that same year. In 2012, Europe as a region imported more oil (617.7 Mt) than the US (524.5 Mt). Together, Europe and

Table 2.2 *Reserves-to-production ratios for coal, oil, and gas reserves*

	Coal (bil. tonnes)		*Natural gas (tcm)*		*Oil (bil. barrels)*	
	Proven reserves	*Recoverable resources*	*Proven reserves*	*Recoverable resources*	*Proven reserves*	*Recoverable resources*
OECD	427	10,657	28	193	244	2,345
Non-OECD	576	10,551	205	597	1,450	3,526
World	1,004	21,208	232	790	1,694	5,871
Share of non-OECD	57%	50%	88%	76%	86%	60%
R/P ratio (years)	132	2,780	71	241	55	189

Source: Based on IEA (2012a) data

the United States accounted for over 41% of all oil imports worldwide (BP 2013). It is useful to point out that among the world's top ten oil-exporting countries, there is not a single EU member, and four of the biggest exporters (Russia, Norway, Mexico, and Canada) are neighbours of either the United States or the European Union.

Another revealing statistic is the geographic distribution of oil refining (see Chapter 1). While oil may originate largely in Russia and the Middle East, almost a third of the world's refining capacity, that is, the ability to turn oil into useable fuel, is located in the United States (18.8%) and the EU27 (16%). Despite being massive producers and exporters of oil, the Russian Federation could only refine around 5,754 thousand barrels a day in 2012, according to BP, or just over 6% of the global capacity, while the Middle East's share of global refining capacity was 8.9%. Together, the OECD's 30 members provide almost half of the world's oil-refining capacity.

The transportation sector relies almost entirely upon petroleum to power vehicles, whether by land, sea, or air, and the fuels they use are produced in refineries that are overwhelmingly controlled by the OECD countries. In 2011, over three-quarters of the total global crude oil-refinery output went to the production of exclusively transport-related fuels (aviation fuel 6.4% and motor gasoline 23.1%), or both transport and heat generation (heavy fuel oil 13.2% and other middle distillates 34.7%) (IEA 2013a). In short, as much as Europe and the rest of the OECD countries may depend on imported oil, crude oil exporters thrive on the synergy provided by consumer-controlled refineries.

The production–export–import picture in the natural gas sector differs slightly to that of crude oil. While natural gas also requires processing, the product-output options are fewer and the ability to store it is more complicated than that of crude oil. As a result, natural gas refinement is much more closely connected to production than crude oil, as it tends to be processed immediately upon capture or shipped directly via pipeline to a processing plant. The United States is the world's top producer of natural gas, and contrary to a few years ago, only ranks sixth among net importers. Amongst the EU member states, only the Netherlands is among both the top producers and top exporters. Four EU member states are among the world's top ten importers, together accounting for over 25% of total world imports of natural gas.

Movements of natural gas are measured in billions of cubic metres (bcm) and are transferred in two ways, either via pipeline as gas or by ship in super-cooled liquefied form, known as liquefied natural gas (LNG). The trade volume of the former (705 bcm) was more than double that of the latter (328 bcm) in 2012. That year, Russia delivered circa 130 bcm via pipeline to Europe, according to BP (2013), constituting the largest single flow of piped gas worldwide.

LNG (not to be confused with compressed natural gas (CNG), natural gas liquids (NGL), liquefied petroleum gas (LPG), or gas to liquids (GTL)) is not a new technology. A commercialized version has been around since the 1960s. However, it is growing in importance. Roughly 327.9 bcm of regasified LNG was traded in 2012 worldwide. The three largest recipients were Japan (118.8 bcm), South Korea (49.7 bcm), and Spain (21.4). Together, these three accounted for over 57.9% of all LNG imports worldwide in 2012 (and because Asian markets pay a premium for the product, they tend to attract the bulk of supplies, accounting for almost 77% of all traded LNG in the year following the Fukushima disaster). Spain's largest suppliers were Nigeria, Qatar, Algeria, and Peru, in that order. Japan received its LNG from Australia, Qatar, and Malaysia. South Korea received its LNG from Qatar, Indonesia, and Oman (BP 2013). LNG technologies are relatively new and rather expensive in comparison to traditional pipeline shipments, requiring advanced ships and seaport terminals fitted with processing technologies to convert LNG back into a gaseous state. Still, new facilities are coming online, and increasingly, advances in LNG technologies are increasing the quantity of LNG suppliers, significantly easing EU plans to diversify its energy sources.

Production and trade movements surrounding the third major fossil fuel, coal, differ significantly from that of oil and natural gas, in that the world's biggest producer is China, which registered a colossal 47.5% of all global coal production in 2012. In contrast, one European country, Poland (1.5%), ranked among that year's top ten suppliers of oil and gas (BP 2013). The United States again mined an enormous amount of hard coal (13.4%), but also ended up as one of the top three exporters, shipping 106 Mt in 2012 (Indonesia 383 Mt, Australia 302 Mt) (IEA 2013a). If coal were still king, Australia and Indonesia would be superpowers. Despite these differences in production and export sources between coal and the other two resources, gas and oil, three EU members (UK, Italy, and Germany) are among the world's largest importers.

Coal continues to play a big role in energy production. The problem with coal, of course, is that it is environmentally destructive to burn and mine. New technologies aimed at sequestering harmful CO_2 may breed new life into coal's future, once again making coal king. For the moment, however, coal is one of the most widely used and yet despised of all primary energy sources.

As the aforementioned figures and data illustrate, the EU is a major importer of almost every primary energy resource, and international competition for those resources is intense. Europe's dependence on foreign fossil fuel supplies creates serious constraints on energy policymakers, who in lieu of finding alternative sources to fuel vehicles, heat homes, and generate power, are forced to limit their international initiatives and economic programmes.

Before closing the discussion on the global trade of energy resources, it is necessary to return briefly to the issue of electricity. Electricity is not traded globally like coal, gas, and oil. This is because physical constraints limit its storage and transmission capacity. However, electricity is traded regularly in thriving regional markets between neighbouring countries, and geography plays a role in determining which sources are used to generate local power. One must also consider that nuclear fuels, and renewable energy sources such as wind, hydro, solar, solar-thermal, and geothermal are also either physically distributed geographically or vary greatly in their local intensity, largely determining the level of their service. Three of these sources (solar-thermal, geothermal, and nuclear power) can generate both heat and power, but all six sources are largely absent from use in the transport sector – the bottleneck being the technology to store electricity generated by these sources to be applied to other end uses. Nevertheless, within the electricity generation sector, contemporary use of two non-fossil fuel sources is noteworthy. These are uranium and hydropower.

Nuclear power accounted for 11% of global and 21% of European electricity generation in 2011. Nuclear reactors use the energy stored in uranium, which is acquired from uranium oxide (U_{308}) mining. According to the World Nuclear Association, around 65,000 tonnes of mined uranium per annum are required to power today's existing reactors. In 2012, new uranium mine production totalled 58,394 tonnes, which equals 86% of world demand, the shortfall being made up from ex-military supplies and stockpiles.

The known planet-wide reserves of uranium are between 6 and 7 million tonnes. Using the aforementioned R/P method, there is enough uranium to supply current usage for another 85 years. However, as in the case of oil and gas, estimates of recoverable reserves tend to change upward, thus R/P ratios remain stable and continue to be subject to debate.

Table 2.3 shows that in 2012, over half of the world's useable uranium came from the combined efforts of Kazakhstan (36.13%), Canada (15.25%), and Australia (11.85%). The trade in nuclear

Table 2.3 *Global uranium production, 2012*

Kazakhstan	36.13%
Canada	15.25%
Australia	11.85%
Niger	7.91%
Namibia	7.62%
Uzbekistan	5.09%
Russia (est.)	4.87%
USA	2.71%
China (est.)	2.54%
Malawi	1.87%
Ukraine (est.)	1.63%
South Africa	0.79%
India (est.)	0.65%
Brazil	0.39%
Czech Rep	0.39%
Romania (est.)	0.15%
Germany	0.085%
Pakistan (est.)	0.076%
France	0.005%

Source: World Nuclear Association (2014)

fuels is tightly controlled by governments and the International Atomic Energy Agency; and for good reason. Uranium is radioactive, harmful to those who come in contact with it, and of course, can be processed and converted into bombs of extraordinary force.

Concluding remarks

Making sense of energy policy requires a basic understanding of its economics. Some noteworthy distinctions, such as between that of a resource and a commodity (and whether the former is finite or renewable), the problems of storability and predictability (both of which are vital for supply and price stability), and the notion of embedded costs and negative externalities, are prerequisites for sound analysis. Equally important is that resources and commodities are used to provide energy services to consumers, whereas consumer behaviour towards energy is driven by the price efficiency of a particular energy service, not the price of a specific energy-carrying resource. Thus, energy economics is driven by derived demand. Consumers do not buy electricity to own it. They buy it to power the devices that make their lives easier.

From the consumer's perspective, energy has three main end uses (electricity, heating, and transportation). While there is flexibility in electricity generation, there is little to none in transportation. Europe has diversified sources and suppliers in recent decades. However, it is still vastly dependent on fossil fuels, most of which lie outside the Union. This creates an uncomfortable vulnerability to price swings and supply interruptions. By diversifying its sources and suppliers (the external dimension), it can mitigate some of those risks. By increasing overall energy efficiency (the internal dimension), it can slow the growth in energy demand. However, even if both policies are implemented to the full, they will have little long-term effect on Europe's comprehensive energy security and sustainability (the multidimensional) unless it has a functioning internal energy market where energy goods and services can be moved and delivered optimally wherever and whenever necessary and profitable.

The energy network, from extraction to delivery, is something like a Rubik's cube. Deciphering the puzzle requires that one always keeps the result (the consumer's use of energy services) in focus, even as one resolves its constituent parts – whether that is responding to externalities from fossil fuel use, increasing the reliability of renewables, or accounting for investment costs required to shift vehicle propulsion away from the internal combustion engine. Just as in the case of a Rubik's cube, solving one side or issue, independent of all others, is a guaranteed path to failure. Understanding the basics of what energy is, how it is produced, and how it is consumed illustrates the interdependent nature of the three dimensions of energy policy as laid out in Chapter 1, and lays bare the many obstacles that states face in achieving comprehensive energy security. Indeed, the concurrent pursuit of security of supply (and price), fair competition, and environmental sustainability sometimes appears to be mutually exclusive, presenting a serious conundrum for policymakers.

It is of course possible to study energy only from the standpoint of resource supply and demand. However, doing so, places the focus of the energy debate on specific energy sources, rather than end-use services. Resource suppliers logically prefer such an approach because it maximizes the value of their exports. The result for policy is that importing countries tend to make substantial short-term investments into research and development to find fuel substitutes, which rapidly taper off when either the price of the commodity drops (because of demand and supply market forces) or because technological efficiency increases (Schubert 2010). Indeed, this is often the unintended consequence of energy efficiency and fuel economy policies and precisely what has happened with recurring investments in alternative fuels for automobiles and efforts to green economies through increased use of renewable electricity.

Finally, the global energy balance matters in all three dimensions of energy policy, because the asymmetric geography of energy resources combined with governments' need to guarantee a continuous and stable supply of energy to their citizens and businesses (so-called security of supply), involves fierce international competition that creates unremitting economic and political tension. It is a business fraught with almost unbridled risk and reward. The fact that negative externalities resulting from energy production, delivery, and consumption know no borders only highlights the inseparable link between national energy policies and international relations.

Nowhere is this complex relationship more evident than in Europe where overall external energy dependency is high, national energy mixes are diverse, and internal market dynamics are being driven by the aim to liberalize and establish one common energy market. Europe of the twenty-first century appears to realize the merits of solidarity in achieving comprehensive energy security, and it has come a long way since founding the European Coal and Steel Community in 1952. Speaking with one voice on energy, however, remains just out of its reach. How Europe has approached integrating the three dimensions of its energy policies and what explains that process is the subject of the next chapter.

Chapter 3

The Changing Nature of EU Energy Policy: Theory and Milestones

The EU's conception of energy policy has transformed over time, reflecting the continuously changing debate of the Union's competences and functions and its theoretical understanding. Given that energy policies are 'product[s] of the interaction of material and technological factors with political ones' (Prontera 2009: 1), its analysis must always be contextualized, that is, put into the right historical context. For example, the supranational–intergovernmental dynamics of integration in the aftermath of its failed attempt to pass a constitution, as well as the dramatic repercussions of the financial crises hitting the Union hard since 2008, and the new tensions between Russia and the EU have played an important role in structuring its energy objectives – member states have simply become more reluctant to transfer more competences to the supranational level while at the same time publicly calling for a new Energy Union. Furthermore, given Europe's dependence on foreign energy sources and their changing form and nature, the global nature of energy questions needs to be considered. Indeed, energy politics is a field where the butterfly effect is not only a common trope but a fact. Embedded into changing global flows of energy, reaching from US shale gas extraction to Canadian tar sands, from Australian LNG to Japanese nuclear power, from the exploration of the Caspian Littoral to Turkish pipeline projects, the Union seems more of an outfielder than a pitcher.

Theoretical perspectives

The history of the Union's energy policy shows a protracted struggle over competences between the supranational institutions, member states, and energy companies driven by business interests. While this may not be different from other policy fields, which slowly drifted towards the supranational realm and were thus 'communitarized' with all the legal implications, energy policy is unique because of the strategic importance the member states attach to it. Building up national champions was the logical step to take, since the constant supply of energy at affordable prices was not yet a question of slowly depleting reserves or environmental concerns. The European Commission's liberalization agenda, which was enshrined in the treaties signed by the member states, had to clash with those national champions. Strategic goods are usually exempted from deregulation. However, the rising environmental agenda and the external shocks showing the finite nature and instability of fossil fuel supply, contributed to a slow opening up of hitherto national energy markets. The lengthy nature of this process can be explained by the existence of powerful national champions, the fact that energy as such is a highly lucrative market with lots of money changing hands and lots of hands being involved (be it lobbies, consumers, or politicians), and the different forms of energy supplies upon which the member states were relying.

The simplest way to look at the changing nature of the EU's energy policy is to look at: (1) the distribution of competences, (2) the overall legislative output over time, and (3) the specific subfields within the energy sector in which legislation was passed (Brutschin 2013).

Contrary to the institutions of the member states, the EU can only act if the member states have transferred competences. This competence transfer is based on the legal principle of conferral. Art. 5 TEU stipulates, 'the Union shall only act within the limits conferred upon by the Member States in the Treaties to attain the objectives set out therein'. As will be shown in the history sections, the EU treaties did not transfer any energy policy competence to the supranational level prior to the Treaty of Lisbon. The term 'energy' itself was only found in three articles (Art. 3, 154, 175 ECT) in the Maastricht Treaty, something which is all the more astonishing given the fact that two of the three EU founding treaties explicitly deal with energy: Euratom and the European Coal and Steel Community. The lack of competence in establishing primary law did not impede the European Commission from initiating European legislation, but it had to rely mainly on regulations pertaining to the establishment of the single market (Art. 114 TFEU), the flexibility clause (Art. 352 TFEU), and general competition law (Art. 101–106 TFEU). It was only in the Treaty of Lisbon that an energy chapter was included and objectives for an integrated energy policy – to be achieved in the spirit of solidarity – were formulated. It is not surprising that those objectives were rather vague. Defining, for example, in which cases the solidarity clause should be applied and which specific mechanism would kick in, would prove to be next to impossible. Without doubt, the energy chapter in the Lisbon Treaty has a political significance that reflects the increased importance of the energy issue for the EU.

The second, simple means by which to assess the changing nature of the EU's energy policy is to look at secondary legislation. The most important legislative acts are regulations and directives. Whereas as regulations are immediately enforceable, directives (usually) have to be transferred into national law (Art. 288 TFEU), that is, directives leave the means to achieve an objective at the discretion of the member state. Figure 3.1 shows the frequency of directives and regulations in the energy sector from 1972 to 2012. It becomes clear that secondary legislation in the energy sector has grown substantially over the years.

In addition, Figure 3.1 shows that from the early 1990s onwards, there was more legislation in the form of directives, allowing more legislative freedom for the member states to achieve the adopted objectives. However, in the wake of the so-called Third Energy Package in 2009, more regulations than directives were passed, signalling that the legislation is getting stricter, leaving member states with almost no legislative leeway.

Given its nature as a strategic good and the huge import dependence of the EU shown during the two oil crises in the 1970s, a closer look at the legislative acts passed in the subfield of energy security shows the changing understanding of what constitutes energy security. In the 1960s and 1970s energy security was mainly understood as security of supply. Over the years, however, two additional dimensions gained importance: competitiveness (affordable prices) and sustainability (environmental issues) (Ciuta 2010). Figure 3.2 shows that until 1996, the majority of the directives and regulations focused on the security of supply, such as restrictions on the imports of crude oil and natural gas. However, since 1998, measures related to the internal market – a tool for increasing competitiveness – and sustainability were on the rise (Brutschin 2013).

Figure 3.1 *Frequency of directives and regulations 1972–2012*

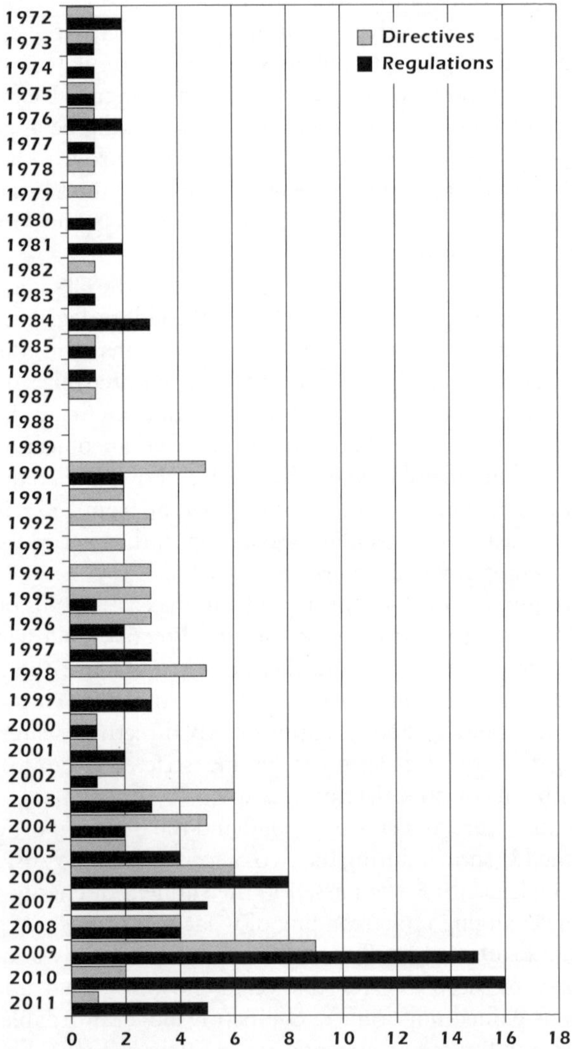

Source: Based on Brutschin (2013: 5)

How can we theoretically frame those conceptual changes reflected in the shift of competences, the number of legislative acts, and the changing substance of regulation? One approach focuses

Figure 3.2 *Directives and regulations by policy area*

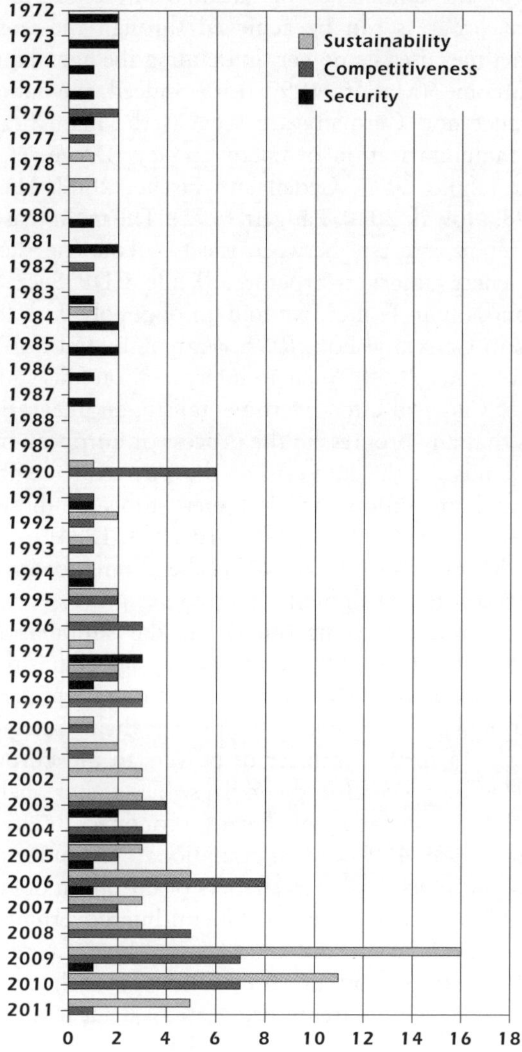

Source: Based on Brutschin (2013: 6)

on member states being the main drivers of the development of the European energy policy. In 1997, Matláry offered an analysis of the period 1985–1995 by applying Putnam's two-level games framework.

She concludes that the deadlocks in European energy policy can be explained by the dominance of national interests. However, she admits that progress can be achieved through the Commission, which sometimes uses its power 'in defining the agenda, the stakes and the outcome' (Matláry 1997: 150). Indeed, the national interests of France and Germany are seen as the main brake to the further communitarization of energy policy (Dreyer et al. 2010, Eising and Jablko 2001, Geden and Fischer 2008, Haase 2009, Lyons 1998, Nowak 2010, Padgett 1992). The main reason for the widely different interests between the two countries lies in their respective energy market structure. While GDF Suez SA has a virtual monopoly in France, we find an oligopoly with four major companies in Germany: Eon, RWE, Vattenfall, and EnBW (comp. Geden and Fischer 2008). Such an interpretation falls squarely into the realm of Classical Intergovernmentalism, an integration theory that posits that any progress in the process of European integration mainly depends on the interests of the powerful member states. Supranational institutions are but mere tools of those member states (Hoffmann 1966, 1982; Milward 1984, 1994).

Andrew Moravcsik's (1993, 1998) Liberal Intergovernmentalism draws attention to the domestic bargaining processes which take place before a policy is negotiated at the supranational level. Governments are agents of this domestically agreed position and try to achieve an outcome closest to their national preferences. Of course, national governments try to optimize their strategies, always conscious that a further transfer of powers to the European level equals a loss of power for themselves. Thus, state behaviour is determined by costs and benefits of economic interdependence (Moravcsik 1993: 480) and supranational institutions serve as facilitators, but much less so as shapers of European policymaking. Matláry (1997) used the frame of Liberal Intergovernmentalism to analyse and explain the outcome of the First Energy Package (1996–1998). It turns out that Moravcsik's approach serves well for the 'grand bargains' (i.e. treaty negotiations), but it is problematic to apply to day-to-day policymaking. The assumption that actors can clearly distinguish between domestic and international levels and that states serve as gatekeepers cannot easily be confirmed. Rather, preferences are shaped at both the European and domestic levels (Schmidt 2000). Under certain conditions the European Commission or interest groups can exert a very strong influence on negotiations.

Thus, European institutions matter. This is also very much the central assumption of theoretical approaches subsumed under the heading of Institutionalism. Rational Choice Institutionalism (Franchino 2007, Tsebelis 2002), similar to Liberal Intergovernmentalism, assumes that preferences are exogenous but that institutions could lead to a 'structure-induced-equilibrium' through certain institutional arrangements, such as decision-making rules (Pollack 2010). According to this approach, both the Commission and the member states shape the outcome of negotiations, but it depends on very specific interest constellations in order for integration to progress. Sociological Institutionalism argues that preferences should be understood endogenously, and that EU institutions can shape them directly, or largely influence them by EU norms and laws (Christiansen et al. 1999: 539). Within the energy context, Rational Choice as well as Sociological Institutionalism have mainly concentrated on the role of the Commission and its legislative powers that can serve as a credible threat to induce more integration (Pollak and Slominski 2011; Schmidt 2001).

Schmidt (1996: 6) argues that an 'orientation toward the dominant theoretical dichotomy blinds one to the multiple effects between supranationalism and intergovernmentalism in what should be more appropriately conceptualized as a system of multilevel governance'. The most important contribution from configuring the EU as a multilevel system is that it allows for a multiplicity of actors, beyond member states and supranational institutions. Interest groups, lobbyists, companies, etc. all try to influence the policy process according to their own preferences. Thus, energy politics becomes a highly complex field in which the interests of supranational, national, and international state and non-state actors are played out. A large number of actors can lead to increased stability according to the veto player theory by Tsebelis (2002). New members (e.g. new member states) may increase the heterogeneity of preferences, but that also means that the status quo is more difficult to change (i.e. stability increases). On the other hand, a larger group of players also eases coalition-building by opening up new possibilities.

A suboptimal policy outcome can be well described by Scharpf's 'joint-decision trap' approach (1988). The European policymaking process is explained not as a result of the widely differing preferences of the member states but as a 'characteristic pattern of policy choices under certain institutional conditions' (Scharpf

1988: 242). The main assumption is that transaction costs grow with the number of actors (Scharpf 2006), and the higher the transaction cost the less likely an agreement becomes. As tempting as the logic of this theoretical approach may seem, Scharpf excludes a central player in his argumentation, namely the European Commission, omitting its role as a key agenda-setter (Scharpf 2006: 850). If the Commission is able to control the flow of information, then adding new members to the negotiating table should not significantly change the speed of decision-making (Brutschin 2013; Kearns et al. 2006). In a revision of his original approach, Scharpf suggests that the 'possibility of "intelligent design" may allow the Commission to present creative proposals that go beyond the trivial exploitation of fixed policy preferences suggested by the role of the agenda-setter in spatial voting theories' (Scharpf 2006: 850).

In the context of energy policy, Intergovernmentalist approaches clearly favour the member states as key actors in the development of the policy field. Whatever the Commission was up to, the member states and, in the case of Liberal Intergovernmentalism, key actors within the member states were acting as the sluice gates through which all Commission proposals had to pass. Institutionalist approaches emphasize the independent role of the supranational institutions, especially the European Commission. Without its constant probing, insisting on the deregulation and liberalization of a highly regulated and protected national field would have never made the progress it has seen over the recent years. Although all the member states were veto players at one point, the significant changes in the primary law of the European Union forced them to live up to their well-trained rhetoric at summits. They are no longer the sole masters of their respective energy fates.

Milestones of European energy policy

The remainder of this chapter explains the evolution of the EU's energy policy from its ambitious and rather successful beginnings to the ambiguity of more contemporary times. It explains how progress towards an integrated European energy policy ebbed and flowed, indicating the crowning achievements as well as the policy failures and the many hiatuses experienced across decades of progress. Instead of looking merely at a single timeline, we approach the development of EU energy policy from the perspective of its path-dependent phases. We demonstrate how the Commission exploited external events in its role as a policy entrepreneur and how the foci of Europe's energy debate vacillated over time between four important themes: the importance given to defining energy resources as strategic commodities at both the national and supranational levels; how and when efforts to deregulate and liberalize various energy markets changed the rules of the game; how the acquisition of sustainable and secure supplies from abroad came to drive many current policies; and how environmental politics arose as an integral part of the Union's energy politics.

We begin by addressing the early phase of European energy policy, from its inception in 1951 to the early 1970s, during which phase Europe focused on securing the basics to both reconstruct its industrial base in the aftermath of World War II and meet its goal to supply European consumers with affordable energy. That process fuelled what has become known as the so-called 'European economic miracle'. It is a phase where the dynamics between the member states clearly dominated the entire integration process. We continue by examining the impact of the two oil crises of the 1970s (1973 and 1979) and the Chernobyl disaster (1986), analysing their role as external shocks triggering further integration. We then turn to the liberalization agenda that dominated EU economic efforts from the early 1990s, before we describe in the final section, the two most recent challenges to an integrated EU energy policy: supply security and climate change. This development of the EU's energy portfolio reflects the major theoretical approaches described above. While in the early years of integration the member states very clearly were the key players, sidelining the High Authority of the ECSC where they deemed necessary, the European Commission later proved very capable in driving the energy portfolio by pointing to the objective of the Common Market – an objective shared by all member states – the increasingly important environmental topics, and today the dominating issue of energy security. Along the way governments, EU institutions, key industrial players, and societal forces were struggling to maintain, increase, or hide their influence on Europe's energy politics. External events very much defined the universe of this policy field. The chapter concludes with an analysis of the EU's 2050 Roadmap and its implications.

The early years: Securing the basics

On 18 April 1951, six European nation states – Germany, France, Italy, The Netherlands, Belgium, and Luxembourg – signed the so-called Treaty of Paris, establishing the European Coal and Steel Community (ECSC). They agreed to abolish all tariff and non-tariff borders in the trade of coal and steel and transfer all authority concerning investment subsidies, price and market regulation, as well as the establishment of a common market to a new committee consisting of nine members, called a High Authority. The ECSC High Authority was the forerunner of today's European Commission. The treaty, which entered into force on 1 February 1953, foresaw a five-year transition phase in which it would continue to allow national subsidies and regulations. Its main objective was not to establish a common energy policy, but rather to integrate German and French heavy industries into a common market, in order to secure cooperative European reconstruction.

Why was so much focus placed on coal and steel? The answer lies in their respective importance to each other. Reconstruction required enormous amounts of steel, and in order to produce steel one needs coking coal. Meanwhile, successful reconstruction would bring about increased demand for electricity, the generation of which in turn required large amounts of boiler coal. The underlying rationale is simple enough. France needed German coal for its steel industry. Germany, whose steel production was booming back to pre-war levels, needed the ECSC to once again become a respected member of the international state community. Both countries needed a great deal of coal and steel to reconstruct, and both materials were regarded as key components of the industry of war. Removing sovereign control over these vital national resources and transferring it to a neutral bureaucratic agency would guarantee mutual economic gain and a peaceful continent. Jean Monnet, one of the founding fathers and the first president of the High Authority, was not alone in hoping that success in one area would spill over into others, an idea conceptualized by functionalist integration theories.

From an economic perspective, the ECSC was only a limited success. Its far-reaching hopes and expectations could not be met. When the unusually warm winter of 1958/59 led to a decline in coal consumption, and therefore to a large surplus, France and Germany reacted in precisely the manner that the ECSC Treaty prohibited, namely by putting up import restrictions to protect their respective national industries. The High Authority simply lacked the power to enforce the treaty. Until the early 1950s, coal occupied an undisputed lead position (almost 90%) in the primary energy supply of the six founding states. By the mid-50s, however, it became clear that the European coal-mining industry could not compete with cheap overseas supplies. It could not stop the meteoric advance of petroleum that was flooding the international energy market either. Consequently, the ECSC Special Council of Ministers and the High Authority openly called for considering a coordinated energy policy (Special Council of Ministers 1957).

In the wake of the 1954 failure to establish the overly ambitious European Defence Community and Political Community, the foreign ministers of the Benelux countries came forward with an important memorandum. The so-called Benelux Memorandum set the tone for the next step of integration by supporting Jean Monnet's idea of an atomic community, one that would provide cheap energy to meet rising demand and thereby reduce costly petroleum imports. Not all the member states were convinced about the merits of such an atomic community, so Jean Monnet suggested linking it with a much broader economic community. Paul Henri Spaak, the Belgian foreign minister (and president of the first Common Assembly of the ECSC), was tasked with hammering out the details of the two communities in an expert group. Despite a multitude of diverging positions, Spaak managed to present a report after only eight months (April 1956), the third part of which singled out energy as an important area of supranational regulation. Progress was sudden and swift. Negotiations between the six ECSC founding members started in May 1956. Most of the opposition to the planned atomic community disappeared after the Suez Crisis in November, which raised doubts about the degree to which Europe could trust oil deliveries from the Middle East. By early 1957, the parties had come to an agreement and on 25 March, signed the Rome Treaties, founding the European Economic Community (EEC) and the European Atomic Community (EAC). Both treaties entered into force on 1 January 1958.

The Euratom treaty had multiple objectives. First, joint research programmes should guarantee an efficient and peaceful use of nuclear energy. For example, the Joint European Torus (JET) agreed upon in 1972, which, later in 2000, became part of the European Fusion Development Agreement (EFDA). Another example is the International Thermonuclear Experimental Reactor (ITER) project, which aims to build the world's largest nuclear fusion reactor in Cadarache, France. Second, it should facilitate developing and implementing common safety standards for the protection of the population. Third, it should provide investment capital for the construction of nuclear power plants, since the costs for doing so were prohibitive; and once those plants were built, they would need to be provided with fissile material in the form of fuel rods. Thus, the fourth objective was to supply such rods, for which a special agency was created in June 1960, known as the Euratom Supply Agency (ESA), which to this day still holds the property rights over all fissile material in the EU and is responsible for its distribution. To ensure that no fissile material is diverted to military use, the member states agreed to install a rigorous control system exercised jointly with the International Atomic Energy Agency (IAEA).

Despite the sense of unity, the member states diverged in terms of their views and expectations about Euratom. One example of this concerned the selection of the reactor model. While France preferred a (somewhat less efficient) model working with naturally enriched uranium, abundant in its Western African ex-colonies, Italy and Germany put their trust in American technology, which functioned with artificially enriched uranium. For France, which was planning its own nuclear military capability, such a dependence on American technology was unthinkable. Meanwhile, the US offered to provide cheap uranium rods and the necessary reactor technology, but rejected the idea of an independent European nuclear arsenal. They demanded that the Europeans abstain from building their own uranium isotope separation plant in exchange for US technology. The argument ended when the US signed a treaty with Euratom to build six nuclear power plants with American money and technology. The final arrangement prompted the then recently elected French President Charles de Gaulle to call Euratom the most infamous treaty France had ever signed.

Euratom was never able to reach the importance of the EEC, even though experts in the 1960s actively discussed atomic energy in terms of a third industrial revolution. The combination of competitive crude oil prices in the early years of Euratom's existence, the division between France and the other member states over whether the nuclear industry should be state-controlled or left to the free market, and clear conflicts of interest between France and the other member states weakened its political importance. A deeper look into the treaty text reveals some of these differences. For example, whereas Article 52 of the treaty stipulated that only the ESA was entitled to buy fissile material, Article 66 provided for special circumstances under which the member states could buy the material independently with the approval of the Commission. Another point concerned joint research projects and the exchange of technical information, where France insisted on the right to far-reaching confidentiality when national security was concerned (see Articles 24–27, 84).

The Suez Crisis in 1956 provided the first, albeit missed, opportunity for the High Authority of the ECSC to extend its competences in the fields of oil, gas, and electricity. Its failure to successfully seize the moment reduced the High Authority's importance almost as fast as coal lost ground as the primary energy source. By 1967 coal accounted for only 35% of total primary energy supply and in 1969, oil replaced coal as the most important source of energy. In short, the ECSC's primary objective to guarantee equal access to cheap coal for all member states lost its relevance. Consequently, the Community gradually began to loosen its prohibition on state aid for the extraction of coal, a step that was deemed necessary to mitigate the negative social consequences of mine closures in the German Ruhr area.

Three crises, all in the Middle East (the 1956 Suez Crisis, the 1967 Six-Day War, and the 1973 October or Yom-Kippur War) revealed the unsustainable depth of member-state dependence on Arab oil, leading them to push the Commission to work out suggestions for the further development of a European energy policy. In response, the Commission prepared a number of important documents (e.g. European Commission 1968, 1972, 1973) in which they envisaged the creation of a common market for energy, including a strategic role for nuclear energy. One outcome resulting from the Suez Crisis, for example, was an agreement to establish strategic reserves, leading to the 1968 directive obliging the member states to maintain minimum stocks of crude oil and petroleum products for 65 days (Council of the European Communities 1968). Beyond this directive, however, the member states could hardly find common ground on any other energy-related issue (Hofer 2008: 39).

Despite the fact that the ECSC, Euratom, and EEC each addressed energy-relevant policy questions and the member states unanimously agreed on their importance, too many obstacles prevented the early creation of a community-wide energy policy.

Already, in its first general report (September 1958), the Commission of the EEC demanded community-wide energy coordination. It subsequently established a working group (25 May 1959) tasked with hammering out suggestions for enhanced coordination and, based on its suggestions, published a detailed report in December 1962 on the future of Europe's energy economy. Only through the tenacious efforts of the supranational institutions, particularly the High Authority and the European Commission, did the member states sign a protocol on an *Agreement on Energy Questions* in April 1964 (Special Council of Ministers 1964), in which they accepted the importance of a common energy market and emphasized their will to create a common European energy policy (Grunwald 2003). The High Authority of the ECSC and the Commission of the EEC were merged on 8 April 1965 in the so-called 'Merger Treaty'.

The development of such a market was negatively affected by a number of conditions. First, energy, due to its specific physical characteristic concerning its production and distribution, is difficult to compare to any other tradable commodity and, thus, requires specific rules and special regulations, which in turn were easier to create at the domestic as opposed to supranational level. Second, energy was (and still is) regarded as a strategic good (Prontera 2009), and because nation states favour national energy champions rather than opt for integrated, transnational approaches, each member state promoted its own vertically integrated utilities. These two factors resulted in the establishment of national energy giants that controlled the entire gamut of the energy industry, from production to transport and delivery. Third, integration was further complicated by the fact that energy is disproportionately subject to technological and scientific advancements. Simply stated, it cannot be predicted when and in which countries scientific and technological breakthroughs will occur. Thus, divergent and sometimes contradictory technological choices were made by member states, which among other things led to different primary fuel mixes and commensurate, sometimes competing, political objectives that years later further complicated the pursuit of integration. Finally, it was very clear from the onset of the European project that the energy sector had a high degree of interdependence with other policy sectors (e.g. competition, external relations, transport, industry, and environment). Therefore, energy policies were often deemed either as a subset of other policies or of such a macro scale that they could not be tackled all at once, further sapping support for full integration of national energy markets because they might negatively impact other key national industries.

Despite the commitment to build a common energy market based on the four freedoms (free movement of goods, persons, services and capital), the EEC treaty was from the very beginning an exercise in the creation of exceptions. For example, one early idea was that there are so-called 'natural monopolies' that should be exempted from competition, such as electricity and gas. Electricity and gas delivery require a physical connection between producer and consumer in the form of transmission or pipelines. If every producer offered its own lines, the costs for energy would explode, while producing energy without having the means to transport it and thereby sell it, holds only limited charm for investors. This is why energy utilities in almost all the member states were both producing and distributing electricity from their foundation. Since investment costs for the construction and maintenance of such grids are considerable, the utility companies were (and still are) reluctant to open their grids to alternative suppliers. There was a strong conviction, particularly in the 1950s, that if the objective was nationwide coverage, there simply was no alternative to vertically integrated utilities; and because those companies were mostly state-owned, it did not contribute to a market-driven approach.

1970–1986: Oil crises and Chernobyl

In January 1973, the United Kingdom, Ireland, and Denmark joined the ECSC, EEC, and Euratom. This was a strategically decisive step for energy supply security in the EU because of the 1960s discovery of huge oil and natural gas deposits in the North Sea. North Sea oil first reached the European market in 1975, reducing its dependence on increasingly unstable Arab suppliers, the degree of which was revealed in 1973/74 when the first oil crisis hit Europe. That crisis occurred in the wake of the October 1973 Egyptian and Syrian invasion of Israel, which Israel managed to repulse with the help of its allies, Europe and the United States.

As a reaction to Western help for Israel, the oil-producing Arab countries – organized through the Organization of Petroleum Exporting Countries (OPEC) – initiated an oil embargo on 4 November 1973. In practical terms, this meant that they cut the production of crude oil. As the EEC imported 95% of its oil at the time and its demand could not change as rapidly, prices soared. A barrel of oil suddenly rose from three to five USD, or an increase of almost 70%. Over the course of 1974, the price rose to more than twelve USD (roughly equivalent to 56 USD in 2014 terms), even though the embargo for most EEC countries had already ended in November 1973 after the Organization of Arab Oil Petroleum Exporting Countries (OAPEC) lifted its embargo on most EEC countries following a joint European declaration on the situation in the Middle East (Statements of the Foreign Ministers and other documents. European Political Cooperation 1973). The sudden increase in cost and growing concerns over supply shortages caused countries such as Germany and Austria to introduce speed limits and car-free days; although the overall economic effect was, in fact, negligible. In the UK, labour disputes with coal miners and railroad workers exacerbated increases in oil prices, leading to a winter energy crisis that forced Prime Minister Ted Heath to ask his countrymen to heat only one room in their homes. The Arab oil suppliers directed their embargo mainly towards the Netherlands, which had allowed the US to use its territory to airlift military supplies to Israel, and provided the latter with oil during the war. European solidarity took a hard hit during the crisis. Despite the high hopes of solidarity raised by the 1968 'stockpiling directive' (Council of the European Communities 1968), the nine member states responded to the first oil crisis by seeking bilateral contracts with Arabian suppliers. Instead of unity, the new attitude was 'save yourself, if you can!'.

The one positive consequence of the oil crisis was the establishment of the International Energy Agency (IEA) in November 1974. The idea behind the IEA was not new. While still National Security Advisor to President Nixon, Henry Kissinger suggested founding an 'Energy NATO' within the institutional framework of the Organisation for Economic Coordination and Development (OECD). The resulting *Agreement on an International Energy Programme*, signed by 16 countries (all EEC member states with the exception of France) in 1974, foresaw an automatic international oil-sharing mechanism in cases of emergency. It then took until 1977 until the member states of the EEC agreed (Council of the European Communities 1977b) to allow the Commission to make suggestions for the Community-wide reduction of oil consumption in cases of emergency. In another missed opportunity, the decision's unclear wording and the lack of any automatic saving procedures very clearly showed that the EEC member states preferred national programmes over European ones.

In September 1974, the EEC Council of Ministers approved a resolution concerning a 'new energy strategy' where they once again emphasized the necessity for a common energy policy and agreed that precise figures were necessary in order to achieve it. They would reach their goal through rational consumption, development of nuclear power, reliance on solid fuels, diversification of supply, technological development, and environmental protection. In mid-December of the same year, a resolution defined the objectives of the Community energy policy to be reached by 1985 (Council of the European Communities 1974). These included reducing Community dependence on imported energy to 50% (down from 63% in 1973), reducing the Community-wide growth rate in energy consumption to a level 15% below January 1973 estimates, and cutting overall petroleum imports from third-party countries to no more than 540 Mtoe (down from 640 Mtoe in 1973). In addition, nuclear power should account for 200 GW of installed capacity; hydro and geothermal electricity contributions should increase to 45 Mtoe. Finally, by 1985, the targets foresaw reducing the percentage of imported oil to 38% of total energy requirements (down from 61% in 1973).

In order to reach its ambitious targets, the Council envisaged voluntary cooperation from industry and the financial support of the member states. The Council passed a few accompanying legal instruments to help the process, including: a regulation concerning the communication on the state of the Community's energy supplies (Council of the European Communities 1976b), a directive regarding a Community procedure for information and consultation on the prices of crude oil and petroleum products in the Community (Council of the European Communities 1976a), and a decision regarding Community procedures for exporting oil and petroleum products between member states in the event of a supply crunch (Council of the European Communities 1977a).

Paradoxically, the oil shortage of the 1970s was offset by a vast surplus of coal. Part of that surplus could be used to implement another Council directive (Council of the European Communities 1975) that obliged member states to maintain a minimum stock of fossil fuels for use in their thermal power stations. Under such circumstances, the ECSC could no longer meet its objective to prevent national subsidies for coal mining. It was now clear that the first oil crises shifted the overriding concerns from the combination of supply security and low prices to secure and sustainable supply.

A second drastic price hike occurred in 1979, triggered by the Iranian Revolution and exacerbated by the onset of the Iran–Iraq war in September 1980. Together, the two events took almost six million barrels of oil off the international market. Coupled with unilateral price hikes by Saudi Arabia and other OPEC nations, the price of crude skyrocketed to more than 35 USD a barrel by 1981 (or circa 112 USD in 2014 terms). By the beginning of 1981, imported oil (excluding freight) cost was 164% more than at the end of 1978, and within the EEC, oil payments rose 40% to 105 billion USD, or almost 4% of EEC GDP. Europe's reaction was limited to the adoption of a general strategy concerning the efficient use of energy and supply security. It comes as no surprise that the European Parliament claimed that it '[r]egrets the constant inability of the Council of Energy Ministers to implement European Council decisions on the common energy policy' (Council of the European Communities 1980: 41).

At the beginning of the 1980s, the Council requested that the Commission provide a progress report concerning the 1990 energy objectives with respect to energy strategy and solid fuels. Increased nuclear capacities and oil production in the North Sea had created a positive supply situation, which allowed the Commission to push for basic agreements in two key areas: energy prices (including taxes) and investments (see European Commission 1980a–d). Harmonizing price levels seemed important to achieve economic convergence across the EEC. Different national policies were leading to enormous price differences between member states, particularly in electricity – in some cases up to 40%. Investing in energy infrastructure seemed an appropriate means to reduce import dependency, so to pay for those investments, the Commission proposed a Community-wide tax on energy.

The 1980 Spring Council in Luxembourg produced precious little in terms of taking up the Commission's latest energy initiative. While it did call for structural changes in the overall use of coal and nuclear power, turning the matter over to the appropriate ministerial councils for further examination, it remained silent on the question of an energy tax (European Commission 1980e). Energy ministers were particularly reserved on the investment question and, thus, commissioned a study of the national energy programmes, with the goal of determining whether such investments were even necessary. In October, the Economic and Financial Affairs Council put under consideration further Commission initiatives on restructuring Europe's energy industries (European Commission 1980f), but the ministers could agree only to postpone the debate over energy prices and investments to a later date. The energy ministers met again in November to clarify the Community's goals in the field of energy, and decided to put the notion of an energy tax squarely on the back-burner. Between 1981 and 1985 the Commission made several energy-related proposals (see e.g. European Commission 1981a–b, 1982, 1983), which the Council politely welcomed and subsequently ignored. Real progress was rare and difficult to identify.

Despite the efforts of the Commission to build an energy infrastructure fund through a community energy tax in the first half of the 1980s, the idea failed to resonate with the member states. Huge differences over state subsidies for coal thwarted any progress. The pattern of indecision was palpable. The European Council would formulate broad policy goals. Then the Commission would hammer out the details and make recommendations. Finally, the member states would fail to agree on any sort of concerted action. In 1983, the Commission issued a progress report in which it concluded that, despite many useful discussions, the Council failed to agree on any measures concerning the Community's energy strategy (European Commission 1983a).

On 26 April 1986 the Chernobyl nuclear reactor in the Soviet Socialist Republic of Ukraine experienced a meltdown and exploded, releasing a massive cloud of radiation that rapidly spread westwards into Europe. The Chernobyl disaster, as it became known, changed Europe's energy focus in the second half of the 1980s. The accident raised deep concerns about the use of nuclear power in general and shattered the popular optimism of the early 1970s that nuclear energy could solve many of Europe's energy problems.

While the political fallout of Chernobyl led to a decisive turn in public opinion about nuclear energy, the actual environmental and health consequences resulting from the radiological exposure were less clear. For example, Cardis et al. (2006) projected that by 2065, radiation exposure from Chernobyl will cause approximately 16,000 cases of thyroid cancer and 25,000 cases of other cancers.

Meanwhile, the UNSCEAR (2000) argued that observed increases in incidents of cancer in the affected areas were observed before the accident (UNSCEAR 2000) and, thus, not necessarily caused by the disaster. In order to 'scientifically clarify the radiological, environmental, and health consequences of the Chernobyl accident', the IAEA initiated the Chernobyl Forum in 2003, which brought together the IAEA with other United Nations Organizations, Belarus, Russia, and the Ukraine. In 2006, the Forum's Environmental Group issued a comprehensive report (IAEA 2006) recommending the continued monitoring of residents in the exposed areas and investigating the long-term effects of radiation on the genetic structure of plants and animals.

The disaster also delivered the uncomfortable reminder that Europe lacked vital early warning mechanisms and public safety measures. Even with a solid treaty base (Art. 51 Euratom), a proper information system and action plan in the case of nuclear disasters was lacking. Thus, the Council passed a Decision on the 'early exchange of information in the event of a radiological emergency' (Council of the European Communities 1987) and the EEC acceded to the IAEA Treaty on information exchange in emergencies. This led to the establishment of an early warning system called the European Community Urgent Radiological Information Exchange (EUCURIE), tasked with constantly monitoring radioactive levels. To address public safety concerns, the Community passed a number of measures. With respect to food contamination, for example, the Council passed a regulation establishing maximum permitted levels of radioactive contamination of feedstock and foodstuffs following a nuclear accident or radiological emergency (Council of the European Communities 1987a). In 1989, it passed a directive on providing public information about health protection measures and steps to take in the event of a radiological emergency (Council of the European Communities 1989).

Almost two months prior to the Chernobyl disaster (17 February 1986), the member states adopted the Single European Act (SEA; 1987), the first major treaty reform since the founding treaties. Best described by the buzzwords deregulation and liberalization, the treaty represented a giant leap towards the completion of the internal market, which the member states agreed to establish by 31 December 1992 and defined it as an 'area without internal frontiers in which the free movement of goods, persons, services and capital is ensured in accordance with the provisions of this Treaty' (Art. 14 SEA). Much of its content was based on a 1985 Commission White Paper entitled *Completing the Internal Market*, which explicitly described 279 separate legislative measures necessary to complete an internal market (European Commission 1985).

In a major reform aimed at facilitating the decision-making process and avoiding delays, the treaty defined an extended number of cases in which Council decisions would require qualified majority voting instead of unanimity, and applied the former to all measures related to the realization of the single market. The treaty also significantly enhanced the role of the European Parliament by introducing the cooperation procedure (Art. 252 SEA) and expanded the EEC's competences in other key policy areas such as research and development, technology, and social policy. Yet the SEA's most important change for the energy sector was the inclusion of environmental law into primary law, which allowed the Community to pass relevant energy laws by relying on environmental law.

Even before the SEA entered into force (September 1986), the member states passed a resolution concerning new Community energy policy objectives for 1995, in which they foresaw the convergence of member-states' policies (Council of the European Communities 1986) and explicitly identified competition as an important mechanism to secure the Community's energy supply, a move which came on the heels of a 1985 Commission White Paper that identified competition as an important mechanism for achieving the internal market (European Commission 1985). This change was an unmistakable shift in favour of a market-oriented approach, and one that stood in marked contrast to the national energy champions that dominated the market at the time. There were several important reasons behind this new approach including the decline of international oil prices, the restructuring of the energy sector that grew out of a general global trend towards deregulation, and the adoption of the Single European Act, which placed the objective of a common energy market squarely in the focus of the EEC's plans to realize its internal market. In keeping with past practice, however, the Council made it clear that it was up to the member states to implement the changes and to secure market principles within their respective economies; supranational guidelines were to serve only as a framework.

A convoluted path from Maastricht to Amsterdam

In September 1986, the Council noted that 'adequate and secure availability of energy' delivered on a 'satisfactory economic basis remains a prerequisite for the pursuit of the economic and social objectives' (Council of the European Communities 1986: 1). In short, cheap and reliable energy is necessary for a strong economy and social welfare. With that principle in mind, the Commission would arduously use the next decade and a half to advance a community-wide energy policy.

During the late 1980s, the Commission exploited the competing interests of the member states in order to foster the building of a common energy market in gas and electricity. In 1988, the Commission published the influential working paper *The Internal Market for Energy* (European Commission 1988), in which it emphasized the connection between supply safety and the internal market. It recommended the strict application of single-market principles to the energy sector. Goods should move freely, monopolies and state subsidies should be banned, and environmental and safety standards should be harmonized. Such a common market for energy was attractive in the eyes of the Commission because most member states were energy importers and their consumers demanded low energy prices. As a first step, therefore, the Commission envisaged a legislative package dealing with areas in which it identified an urgent need for liberalization: gas and electricity transit, price transparency, and investment projects.

Most member-state governments were initially sceptical. Germany and the Netherlands demanded a broader European approach. Denmark was strictly against the introduction of market principles due to its opposition to nuclear energy, and a free market would make it next to impossible to control the sources of electricity production. Spain was afraid to lose profitable income earned from transit charges levied on electricity transfers between France and Portugal. Greece worried about domestic coal production and its chances to survive in a free market. Furthermore, the energy industry voiced strong opposition to the gas transit proposals. The two big exceptions were France and the UK. The former wanted to export its substantially overproduced nuclear-generated electricity, and the latter had already finished liberalizing its national energy market in the preceding years.

The so-called common carrier principle, where owners would be required to open existing gas and electricity grids to alternative suppliers in exchange for a fee, and consumers were given the freedom to choose between energy suppliers, proved to be the most contentious points. Therefore, in September 1989, the Commission installed two committees to discuss those questions. The Professional Consultative Committee on Electricity (PCCE) brought together representatives of the electricity-generating companies, and the Comité Consultatif Etats Membres Electricité (CCME) brought together the twelve member states. It took four and a half years of protracted discussion before a common directive (EP/Council 1996) was passed in which the member states agreed *inter alia* to vertical unbundling and increased competition including a phased opening of supply competition, but it failed to provide for effective regulation and avoided the issue of granting monopoly rights. In essence, it was a first step, a pragmatic exercise in balancing collective interests.

Both internal and external reasons explain the lengthy negotiations, and the less than optimal outcome. Internally, the member states needed to overcome a fundamental contradiction present in their energy markets. Europe's energy-intensive industrial sectors wanted access to France's nuclear-generated electricity because it was much cheaper than German electricity, which was being produced by coal-fired power plants. However, allowing this would reveal the enormous subsidies that both states were pouring into their national champions. Meanwhile, rising dependence on foreign electricity raised concerns by these same national champions about their legal and political responsibility to provide a constant and stable supply of energy, not just for big industry but also for the public at large.

Once again playing the role as agenda-setter, the Commission tried to overcome the respective national opposition by formulating a variety of legislative proposals with enough integrated leeway to account for the competing interests of the member states. It was a strategy that achieved limited success, both in terms of legislative passage and its publicly declared internal market goals. For example, while a draft directive concerning the mutual notification on investments in the oil, gas, and electricity sector was rejected by the member states because they deemed it to be too bureaucratic, the Council approved, by qualified majority, a directive on the transit of natural gas through grids (Council of the European Communities 1991) that intended to break up the distribution monopoly of the large utility companies. This came on the heels of another, at least partial, legislative success, namely a draft directive to improve the transparency of gas and electricity prices charged to industrial end users (Council of the European Communities 1990a). The latter deliberately omitted discussion of prices in order to allow for easy adoption. The Commission also tried to put pressure on the member states to finally abolish state subsidies, criticizing subsidies for coal mining in Spain and Germany as well as in the petroleum sector in Spain, Portugal, and Greece. Alas, such massive structural changes in the subsidy regime proved to be too much, too fast for the member states.

Externally, a series of events occurring over the course of the negotiations substantially changed the political and energy landscape. Among these were the fall of the Berlin Wall (November 1989) and the subsequent German reunification (October 1990), the Iraqi invasion of Kuwait (August 1990) and the subsequent Gulf War (February 1991), and the tumultuous dissolution of the Soviet Union (December 1991). For some Western European leaders, the fall of the Iron Curtain evoked a long-held dream to unify the continent under the auspices of the European Community. The necessary preparations for taking in new members from the East were reflected in the Treaty on the European Union (TEU) signed on 7 February 1992 in Maastricht in the Netherlands. When the TEU ('Maastricht Treaty') entered into force on 1 November 1993, it was the first to explicitly mention energy in primary law, albeit not in the form of an energy chapter as sought by the Commission. The member states could only agree at that time to include energy in the list of activities and objectives (Art. 3 TEC). Contrary to other policy fields (e.g. Common Agricultural Policy), in which specific and concrete instruments for the achievement of an objective were included, no measures were taken for energy policy. Following the same pattern, the Treaty of Amsterdam did not touch on energy policy at all. It was only in the Constitutional Treaty, which failed ratification in 2005, that a coherent energy article (Art. III-256) appeared for the first time.

Only three provisions of the TEC mentioned the term 'energy'; and they did so in either general terms or broad goals. According to Article 3(1) TEC, the activities of the Community should include 'measures in the spheres of energy'. Similarly, according to article 154 TEC, 'the Community shall contribute to the establishment and development of trans-European networks in the areas of transport, telecommunications and energy infrastructures'. With regard to environmental policy, Article 175(2) TEC stated 'the Council [...] shall adopt [...] measures significantly affecting a member state's choice between different energy sources and the general structure of its energy supply'. For most of its existence, therefore, European energy policy relied on a mixture of general internal market rules (Art. 114 TFEU, ex Art. 95 TEC), the application of the flexibility clause (Art. 352 TFEU, ex Art. 308 TEC), and environmental and competition law (Art. 101–106 TFEU, ex Art. 81–86 TEC). Competition law was particularly important in the pre-Nice Treaty years, because it dealt with mergers and acquisitions, antitrust laws, and state aid provisions, areas that strongly relate to the energy sector.

Increasing political instability in Russia raised serious concerns about the security of Europe's gas supplies, prompting Dutch Prime Minister Ruud Lubbers to suggest establishing a European Energy Charter during the June 1990 European Council meeting in Dublin. The Council responded by asking the Commission to draw up a charter particularly concerning the production and transport of gas and to present it at a special energy conference in the fall of 1991. An attempted coup in Moscow (August 1991) accelerated the process. In December 1991, the member states, meeting in The Hague together with representatives of the European Free Trade Association, the Central and Eastern European accession states, Japan, and the US, signed the European Energy Charter. The Charter itself, formally known as *Concluding Document of The Hague Conference on the European Energy Charter*, was a 'concise expression of the principles that should underpin international energy cooperation' mainly as it related to securing energy supplies (see Chapter 7).

Promoting international norms in the energy sector was one of the first attempts by the European Community to employ a strategy of rule export in order to secure its interests abroad. For this reason, among others, the 1991 agreement was deliberately non-binding, with the explicit understanding that the signatories would work towards a binding multilateral agreement (Energy Charter Secretariat 2004). Negotiations began promptly and the Energy Charter Treaty was signed in Lisbon in December 1994. Once in force (1998), energy was to be traded according to GATT/WTO rules. Foreign investments were to be protected. Exploration, production, and transport were to be non-discriminatory, and energy transit was to be guaranteed even in times of conflict. Russia signed the treaty, but never ratified it, and in 2009 officially announced that it will not become a contracting partner (although it remained a member of the Energy Charter Conference); the practical effect being that foreign direct investment in Russia's energy was no longer protected under Charter rules. The US limited its membership to observer status, as did Canada; and Norway had yet to ratify its membership in late 2014. Nevertheless, by January 2014, 52 states had signed or ratified the treaty, most of which were located in Europe or Eurasia. Thus, although the strategy of rule export in order to increase supply safety failed to sway its biggest external energy supplier, Russia, the European Community did succeed in setting the agenda and the rules for much of the former Soviet Union, and, in that important sense, was quite successful.

The late 1980s and early 1990s witnessed fundamental changes in Europe's political and energy landscape. It had taken important steps towards an internal energy market, capitalized on external circumstances to enhance its treaties, and begun the process of establishing advantageous norms in the international energy market. Concurrent with these developments was a growing awareness of environmental matters.

Environment seizes the day

In June 1988, over 300 scientists and world leaders met in Toronto for the first World Climate Conference. Known officially as Conference on the Changing Atmosphere, they established a clear link between energy consumption and climate change, focusing international attention on the issue. A year later, the participants of the 14th World Energy Conference in Montreal concluded that while rising global energy needs could be satisfied technologically, the resulting environmental damage would constitute a serious problem. The conference was organised by the World Energy Council (WEC), an institution formed in 1923 establishing a network of leaders and practitioners to exchange ideas and concepts on how to create a sustainable energy system, and which, by 2014, had united over 3,000 member organizations from a variety of fields. Despite the dire warnings and growing body of evidence, however, the wealthy industrialized countries were unable to reach a consensus on how and to what extent greenhouse gas (GHG) emissions had to be reduced in order to avoid the worst-case scenarios.

Out of this hotly debated international topic, the European Commission emerged as a driving force. In February 1990, the Commission published guidelines for an environment-friendly energy policy and asked the member states to take measures to reduce industrial emissions (European Commission 1990). Two months later, the Council and Parliament passed a regulation establishing the European Environment Agency (EEA), which only began operating in 1994, and the European Environment Information and Observation Network (Council of the European Communities 1990b) to support the Community and member states on environmental issues and improve cooperation between them. At the second World Climate Conference in Geneva in 1990, the Community advocated adopting measures against climate change, perhaps as a response to the dramatic environmental damage found in some areas of the former Eastern bloc countries following the fall of the Iron Curtain. China, Japan, the Soviet Union, and the United States did not agree. In fact, the parties could only agree on the future coordination of activities and the intention to develop a convention on climate change. The World Meteorological Organization (WMO) provides a summary overview of the World Climate Conferences on its website (wmo.int).

In response to these developments, Europe's environmental and energy affairs ministers held a joint Council meeting in October 1990, the first of its kind, where they agreed to stabilize CO_2 emissions at 1990 levels by 2000. Against this backdrop, the Commission proposed an ambitious directive in June 1992 aiming to reduce CO_2 emissions by once again reviving the unpopular notion of an energy tax, specifically one that targeted petroleum fuels and electricity (European Commission 1992). Anticipating resistance, the Commission formulated the draft cautiously, proposing special conditions for energy-intensive industries and encouraging their compliance through tax rebates on investments in energy-saving technology. In order to reduce possible competitive disadvantages, it also proposed making the introduction of such a tax dependent upon other OECD members introducing similar measures. The latter was a deal killer. Besides the obvious problem that the member states would never agree to it, it was clear that the other OECD countries were not even considering introducing such a tax. The proposal can be seen, therefore, as the Commission's testing of the political waters. Another interpretation, however, is that the Commission was attempting to establish a progressive position in light of the 1992 UN Conference on Environment and Development (UNCED) held in June in Rio de Janeiro, and one that fit nicely into its publicly crafted image of environmental advocate.

Two and a half years later (December 1994), a revised proposal was also shot down at the Essen Council meeting, but not because of a principle objection to energy taxes as such. Ultimately, competing interests between member states killed the bill. Coal-consuming countries wanted the tax applied to all energy sources, while nuclear and natural gas consumers were only prepared to accept a tax on CO_2 emissions. The gap was too large to bridge. Even a compromise would have been untenable, because the member states were openly reluctant to grant the newly formed European Union any taxation authority or competence.

Despite the failed attempts to establish a Community-wide energy tax and pass overly ambitious environmental legislation, environmental and energy policy became permanently linked by the time the European Union came into force. Since then, the reduction of CO_2 emissions, investments in renewable energy production, and increased energy efficiency, etc. have become the corollary – and sometimes even the source – of EU energy politics. Indeed, environment has seized the day.

When in 1991 the European Parliament symbolically declared 1994 to be the year of energy (European Parliament 1994), the Commission welcomed the initiative. It saw the EP as a partner to further its own energy policies and seized the opportunity. In February 1995, just a month prior to the first UN World Climate Conference in Berlin, the Commission published a Green Paper entitled *For a European Union Energy Policy* (European Commission 1995). This was followed by a White Paper in December 1995 entitled *An Energy Policy for the European Union* (European Commission 1995a) that emphasized the complete integration of energy policy into the general economic policy, which included market integration, deregulation, limits on public intervention, sustainable development, consumer protection, and economic and social cohesion. According to the Commission, the goal of the European energy policy was to achieve competitiveness, supply security, and environmental protection, something that it would later call comprehensive energy security (see Chapter 1). To emphasize these goals, it established the Energy Consultative Committee in 1996 and gave it a five-year mandate. The Commission subsequently announced its intention to increase the market share of renewables up to 12% by 2010 (European Commission 1996, 1997) and later transformed the Energy Consultative Committee into the European Energy and Transport Forum, comprising 34

Commission-selected members from diverse sectors involved in energy and transport policy. It also limited its mandate to renewable two-year terms (European Commission 2001c).

Meanwhile, further incentive for action came from the growing international momentum to curb global warming. In December 1997, the United Nations Framework Convention on Climate Change (UNFCCC) held a climate change conference in Kyoto, Japan, resulting in an international treaty that aimed to reduce growth in greenhouse gas emissions. The resulting and legally binding Kyoto Protocol, which the European Union signed in April 1998, contained specific emissions targets for the so-called Annex I countries. Countries with commitments under the Kyoto Protocol were expected to meet their obligations to reduce greenhouse gas emissions primarily through national measures. Accordingly, the EU agreed to reduce its CO_2 emissions by 8% by 2012, compared to 1990. In order to achieve this ambitious goal, the Commission proposed to double the share of renewable energy (from 6% to 12%) in total energy consumption by 2010. In short, they could now pursue the recommendations made in the 1997 White Paper (European Commission 1997). The Commission further proposed increasing the share of biofuels to 5.75% and increasing energy efficiency annually by 1% by 2010. However, implementation was never very likely because it lacked appropriate sanction mechanisms. Nevertheless, the Commission continued to stay on message in the years that followed. For example, in October 1998 it published *Strengthening Environmental Integration within Community Energy Policy* (European Commission 1998) where it pushed its view that environmental protection and an efficient energy policy were compatible notions, following up in April 2000 with an action plan to reduce energy consumption through improved energy efficiency (European Commission 2000a).

The coming into force of the Maastricht Treaty provided the necessary boost for the Commission's liberalization efforts aimed at increasing competition, keeping energy prices low, and securing existing and future energy supplies. Several important legislative acts issued in the early 1990s paved the way for those efforts (see also Chapter 5, Section 'The creation of the internal energy market'). For example, the Council issued three directives by late 1991 that improved the transparency of gas and electricity prices for industrial users (Council of the European Communities 1990a), facilitated the transit of electricity between the high-voltage grids

(Council of the European Communities 1990c), as well as the transit of natural gas between high-pressure transmission grids (Council of the European Communities 1991). Focus on the electricity and gas markets continued into the second half of the 1990s when the European Parliament and the Council issued two more directives for electricity and gas (EP/Council 1996, 1998) establishing common rules for their respective internal markets, with a special focus on simultaneously increasing efficiency in production, transmission, and distribution, and 'reinforcing security of supply and the competitiveness of the European economy [while] respecting environmental protection' (EP/Council 1996). In an illustration of how much the European Union had grown since the mid-80s, both documents stipulated concrete goals to be achieved in the immediate future, while at the same time guaranteeing that member states retained enough room for manoeuvre.

One can observe that the liberalization efforts of the 1990s delivered mixed results. There was progress in the natural gas sector, but it was slow and protracted. Transparency and cooperation, and the beginnings of renewable targets, in the electricity sector laid the foundation for achieving the European Council's goals to establish trans-European gas and electricity networks. However, the opening of national electricity markets suffered under the tension between competition and comprehensive provisioning; and the development of trans-European gas and electricity networks stagnated due to insufficient funding. Nevertheless, the decade brought agreements such as Directive 96/92/EC that required the member states to open their national electricity markets to foreign competition; and once it entered into force in 1999, there was no going back. The most remarkable development, though, concerns the position environmental affairs took in the realm of energy politics. It became the guiding tool for Europe's energy policy – if not always in fact, then certainly in rhetoric.

Supply security and climate change in an integrated market: From Nice to Lisbon

In response to the sharp rise in the price of oil in September 2000, the Commission issued the Green Paper *Towards a European Strategy for the Security of Energy Supply* (European Commission 2000), which called for an integrated approach to promote energy self-sufficiency, environmental protection, and market integration.

The Commission described the EU's energy situation in the bleakest of terms, identifying its expanding external dependence as creating 'economic, social, ecological and physical risks', and noted that the EU lacked the 'necessary means to change the international market', a weakness laid bare by the spike in oil prices. According to the Commission's online summary statement, 'The European Union's (EU) external energy dependence is constantly increasing. The EU meets 50% of its energy needs through imports and, if no action is taken, this will increase to 70% by 2020 or 2030.' Those energy imports, it adds, 'account for 6% of total imports and, in geopolitical terms, 45% of oil imports come from the Middle East and 40% of natural gas imports come from Russia' (European Commission 2014m).

The Green Paper called for a broad public discussion to be completed by November 2001. Among the issues to be discussed were ways to reduce energy import dependency, the role of taxation and public subsidies in the energy sector, funding strategies for renewables, supply grid optimization, the future role of nuclear power, measures to fight climate change, and ways to save energy. Most of the member states and NGOs that commented upon the Green Paper welcomed the Commission's proposal to control energy demand by promoting energy efficiency. Since renewables could be produced at home, their increased use should reduce import dependency and their higher cost could be offset with savings through energy efficiency. The intensive discussions incited by the Green Paper were summarized later in a Communication from the Commission to the EP and the Council (European Commission 2002a). In February 2002, the Commission formulated the Communication *Towards a Global Partnership for Sustainable Development* (European Commission 2002b), in which it explained how the Union could contribute to sustainable global development and identified key components that it recommended to be incorporated into an international agreement, a suggestion taken up by the 2002 Barcelona European Council. In addition to a White Paper on transport policy (European Commission 2001), a tremendously important sector that is responsible for 32% of the EU's energy consumption and 28% of all its CO_2 emissions, two important pieces of legislation followed: a directive on electricity generation from renewable energy sources (EP/Council 2001) and one on the promotion of biofuels for transport (EP/Council 2003a).

Despite the tendency towards a stronger focus on renewable energy sources, the Treaty of Nice, which entered into force in February 2003, brought about no discernable progress in the energy sector. Once again, the member states were simply unable to reach a consensus concerning a common energy policy. That was reserved for the Convention on the Future of Europe where a select group of representatives of national heads of state, parliaments, the EP, and the Commission, under the leadership of Giscard d'Estaing, were tasked with coming up with proposals to reform the EU's political system. The Convention met between 28 February 2002 and 13 June 2003. Once it began, however, it became clear that the body would not limit itself to the mere drawing up of proposals. Rather, it aimed to draft a European Constitution. The resulting *Draft Treaty Establishing a Constitution for Europe* (European Convention 2003) contained a major breakthrough for EU energy policy. It proposed shared competence in the field of energy (Part I, Title III, Article 13, Paragraph 2) and explicitly included a new section on energy (Art. III-157). It also set well-known goals, including establishing a functioning common energy market, achieving energy supply security and increased energy efficiency, as well as increasing the use of, and developing new, renewable energy sources. The delegates planned to substantiate those goals with laws or framework legislation after consulting with the Committee of the Regions and the European Economic and Social Committee. In order to reduce expected resistance, the section emphasized state autonomy in terms of their energy mix, writing '[s]uch laws or framework laws shall not affect a Member State's choice between different energy sources and the general structure of its energy supply' (European Convention 2003: 57).

The draft constitution had to overcome two major hurdles. First, it required ratification by an intergovernmental conference, which is the only institution that is authorized to change primary law according to the Treaties (Art. 48 TEU), and ratification by the member states in accordance with their national laws. Ultimately, the intergovernmental conference signed the Constitutional Treaty in October 2004, although not without at least one failed attempt under the Italian presidency of the Council in December 2003. Intriguingly, the energy article survived the debate unscathed (Art. III-256 TCE). The second hurdle, however, proved to be much more difficult. In a May 2005 referendum, a majority of French voters rejected the constitution. The same happened in the Netherlands in

June. As a consequence, when the European Council met later that month, they decided to launch a 'period of reflection' on the future of Europe (European Council 2005). The constitution for Europe was dead in the water.

It took until June 2007, when another intergovernmental conference was convened, to develop a 'Reform Treaty' in order to revise the existing EU treaties. The heads of states and governments signed the *Treaty Amending the Treaty on European Union and the Treaty Establishing the European Community* in Lisbon (Treaty of Lisbon or Lisbon Treaty) on 13 December 2007. The energy section from the Draft Constitutional Treaty was largely copied into the new Lisbon Treaty, with one major exception – Poland insisted on a solidarity clause with respect to energy (Art. 194 Para. 1 TFEU). Among the objectives were the usual suspects, but the article contained a new objective concerning the promotion of the interconnection of energy networks. This additional objective was important because sufficient transmission capacity between member states is a prerequisite for a functioning internal energy market. Such networks would greatly reduce the almost total external energy dependence of some Eastern member states on Russian gas. Another decisive treaty element concerned the application of the ordinary legislative procedure (previously known as the co-decision procedure) to the field of energy. The ordinary legislative procedure replaced the co-decision procedure when the Lisbon Treaty came into force (Article 294 TFEU), providing the European Parliament with equal weight as the Council on a wide range of areas, including energy. Thus, although member states continue to retain full authority over their respective energy mixes, the Lisbon Treaty marked a substantial leap forward for an integrated European energy policy.

Article 194 TFEU formulates the following energy policy goals:

- to ensure the functioning of the energy market;
- to guarantee energy supply security in the EU;
- to promote energy efficiency, energy savings, and the development of renewable energy sources;
- and to promote the interconnection of energy networks.

These goals are to be realized 'in the context of the establishment and functioning of the internal market [...] with regard for the need to preserve and improve the environment [...] in a spirit of solidarity between Member States' (Art. 194 TFEU). The measures taken to meet these goals are adopted in accordance with the so-called 'ordinary legislative procedure', meaning that the majority of representatives in both the Council and in the EP must support the measure. Nevertheless, if a decision has an impact on a member states' energy mix and its supply security, the Council still has to decide unanimously. This would be the case, for example, if the Commission initiated a EU-wide withdrawal from nuclear energy. Unanimity is also required if a proposal primarily touches existing tax legislation, in which case the EP only possesses the right to be heard.

Listing the different objectives of the EU's energy policy in Article 194 TFEU does not mean that those objectives are equally important, or that they can be achieved concurrently. In fact, much tension exists between those objectives, as do the conflicting goals of member states. For example, every member state is granted the right to determine the 'the general structure of its energy supply' (Art. 194(2) TFEU). However, this may conflict with the EU's target to establish a functioning internal energy market and promote the interconnection of energy networks as laid down in Article 194(1) TFEU. Similarly, a member state's right to determine 'its choice between different energy sources' (Art. 194(2) TFEU) – the specific energy mix – may well offset the EU's climate change and energy efficiency targets. Thus, despite Lisbon's significant upgrade in the legal basis and processes to meet its energy goals, the law remains vague on precisely which mechanisms can be applied to ensure solidarity between the member states in the event of an energy crisis or supply disruption.

Independent of the changes in primary law, the Commission furthered its goal to establish a common energy market in the first few years of the twenty-first century. In June 2003, the Council and EP adopted directives to establish common rules for the internal market in electricity (EP/Council 2003b) and to open national gas markets (EP/Council 2003c). The main purposes of these directives were to unbundle transmission and distribution system operators and their accounts, and open access for third parties to transmission and distribution systems. Additional steps were taken to enhance 'conditions for access to the network for cross-border exchanges in electricity' (EP/Council 2003). As the Community's competition rules prohibited discrimination, the member states were requested to ensure that third parties could access transmission and distribution systems based on tariffs universally applied to all eligible customers. The Council, the EP, and the Commission then took a number of additional steps to maintain progress and enhance harmonization. For example, independent national regulatory authorities, each of which varied in its responsibilities, still oversaw national market conditions and practices. Thus, the Commission established the European Regulatory Group for Electricity and Gas (European Commission 2003) to advise it and national regulators. In the environmental field, the European Parliament and Council passed Directive 2003/87/EC establishing a trading scheme for a greenhouse gas emission allowance, which came into force on 1 January 2005 (EP/Council 2003d; European Parliament and Council of the European Union 2009c).

In January 2007, the Commission published two important Communications with very different and yet complementary messages. In its *Sustainable Power Generation from Fossil Fuels: Aiming for Near-Zero Emissions from Coal after 2020* (European Commission 2007l), it acknowledged the importance of fossil fuels. In its *Renewable Energy Road Map*, it set out a long-term vision for renewable energy sources in the EU and proposed to 'establish a mandatory (legally binding) target of 20% for renewable energy's share of energy consumption in the EU by 2020' (European Commission 2007m). The reasons behind this highly ambitious goal were the events of winter 2004/05, which drastically revealed Europe's vulnerability in the energy sector.

A series of energy supply crises between 2006 and 2009 once again reminded Europe of its unfavourable predicament vis-à-vis external suppliers. When Russia suspended all gas supplies passing through Ukrainian territory on 1 January 2006, there was a sudden and sharp drop in gas pressure along the Austrian, Italian, Polish, and German gas lines. The crisis occurred as a result of a dispute over prices and debt. While Gazprom demanded that Ukraine pay a market rather than preferential price for its natural gas, Ukraine demanded higher gas transit fees. Almost 80% of the gas Russia exported at the time to the West transited Ukraine. Consequently, a number of European countries experienced sharp drops in their supply.

The EU reacted by passing a directive concerning 'measures to safeguard security of electricity supply and infrastructure investment' (EP/Council 2005), with the objective to reach supply security via a competitive single EU electricity market and joint investments in infrastructure. Shortly after the crisis abated, Poland openly proposed establishing a European Energy Security Treaty (Council 2006a), a sort of 'Energy NATO' alliance with far-reaching goals beyond what Kissinger had envisaged in 1975 (Scharples 2012). While the proposal failed to take hold in 2006, it is notable that some progress on establishing an Energy Union came under a 2015 Polish presidency of the European Council. Nevertheless, the quarrelling between Russia and Ukraine repeated itself in the winter of 2007/08, and again in 2008/09, when Transneft, another Russian state-controlled business responsible for its national oil pipelines, decided to close the Druzhba (Friendship) pipeline, which back then, delivered 20% of German oil imports. It took three days of intensive negotiations to reopen the pipeline. The disputes between Russia and Ukraine reached a crescendo in 2014 when Russia seized the Crimea in the wake of a pro-European revolution, which itself followed the sudden about-face by Kiev in the negotiations over a European association agreement. This set off a violent uprising by pro-Russian separatists in eastern Ukraine, which put Brussels and Moscow on a collision course that resulted in a major deterioration of EU–Russian relations and the suspension of the South Stream pipeline project. At the time of writing (May 2015), the conflict remained unresolved.

For Europe, the Russian–Ukrainian crises of 2006 to 2009, and especially the events of 2014, left a sour taste, not only because it showed the extent of its dependence on Russia but also because it revealed another vulnerability in its dependence on neighbouring transit countries, and exposed a number of fundamental weaknesses in Europe's energy situation. It was well known that Europe was far too dependent on foreign gas supplies, but Europe's inability to move surplus gas and oil supplies to its newer member states in need, particularly in the 2005 and 2008/09 cases, underscored the problems of not having a fully integrated internal gas market. Russia, which was equally vexed because it relies heavily on receipts from its deliveries to Europe, ramped up its efforts to find ways to supply Europe by bypassing Ukraine altogether. The latest crisis came at an opportune time for Moscow. Russia had already proposed to build a pipeline to bypass Ukraine altogether in 2001, by moving offshore directly from Vyborg (Russia) to Greifswald (Germany), and had been conducting feasibility studies and securing the requisite permissions ever since. The result was the construction of the Nord Stream pipeline consisting of two parallel lines, the first of which became operational in November 2011 and the second in August 2012. Europe thus came out of these crises no less dependent on Russian oil and gas.

In March 2007, the Commission, seeking to stimulate debate, launched another Green Paper concerning market-based instruments for environment and related energy policies, and once again included the notion of an energy tax (European Commission 2007c). Later that month, the Council adopted an Energy Action Plan, which included, among other things, a reduction of European emissions by 20% compared with 1990 levels (European Commission 2007). Similar targets were also included in the *20-20 by 2020* strategy the Commission presented in 2008. The easy to remember slogan implied two important goals: first, that the EU has to reduce greenhouse gas emissions by at least 20% and, second, that the member states have to increase the share of renewable energy to 20% of total energy consumption by 2020 (European Commission 2008a).

Then in 2009, the Commission published a progress evaluation of the so-called second legislative package on energy (the first consisting respectively of the 1996 electricity and 1998 gas directives). It concluded that even four years after the deadline (1 July 2004), implementation of the second electricity and gas directives was still incomplete (European Commission 2009a). The Commission complained about the member states' compliance record, already noting in 2007 that 'the objectives of market opening have not yet been achieved. Despite the liberalisation of the internal energy market, barriers to free competition remain' (European Commission 2006a). The European Regulators' Group for Electricity and Gas (ERGEG), which was responsible for monitoring the progress of the member states, reported that some were still far from full compliance with the directives (European Commission 2009a). Further shortcomings existed in numerous areas. For example, gas and electricity wholesale and retail markets were still highly concentrated. Gas and electricity prices varied substantially between member states. Unbundling measures were not sufficiently implemented. Information on consumers' rights and choices was lacking. Worst of all, the EU was still inadequately prepared to react to a potential gas supply crisis (European Commission 2009a).

The Commission initiated a third legislative package based on these insights, which, proposed in 2007, entered into force in September 2009. As in previous years, two directives intended to remove the remaining barriers were simultaneously adopted for the gas and electricity markets (EP/Council 2009a, b). The member states were given 18 months (until March 2011) to implement them. This new package aimed at increasing access by third parties, transferring competences to the Agency for the Cooperation of Energy Regulators (ACER; founded in 2010 – an institution that was expected to assist the national regulatory authorities and eventually coordinate the work between them), and breaking the market stranglehold held by vertically integrated companies. The latter point was prioritized because the strong and intransigent position held by vertically integrated energy companies continued to pose a major challenge to the liberalization of the member states' gas and electricity markets. Therefore, while former directives regarded merely the legal unbundling of generation and transmission as sufficient, the 2009 legislative package (the so-called third legislative package) added the feature of separating companies' generation and sale operations from their transmission networks, known as ownership unbundling. This piece of legislation was designed to further liberalize the internal energy market. However, it brought about external repercussions for EU relations with Russia, because it excluded the possibility of state-owned companies, such as Gazprom, to move gas into Europe and concurrently own the transmission and sale of gas. This problem ultimately played a part in debunking Russian support for the South Stream pipeline project and heightening tensions between Moscow and Brussels.

The EU was no less active on the regional and global levels throughout the early years of the twenty-first century, as it actively pursued a strategy to build deeper cooperation with a wide range of countries, notwithstanding its dispute with Russia over the fate of Ukraine and the Crimea. For example, it commenced the EU–Russia energy dialogue in 2000, initiated a new Mediterranean aid program for the years up to 2013 that aimed to integrate the European and Maghrebi gas markets, and extended the EU energy acquis to the Balkans through the Energy Community of South East Europe (ECSEE). It also incorporated Azerbaijan into its Neighbourhood Policy to capitalize on the country's important energy resources, an arrangement that paid off in December 2013 when a European consortium of companies signed a deal to develop the massive Shah Deniz 2 gas field just as Ukraine, under pressure from Moscow, withdrew from negotiations to upgrade cooperation with the EU.

Similarly, the EU increased its focus on regional partnerships, including Africa (the Africa–EU Energy Alliance) and in the Black Sea region. The Baku initiative, for example, brings the Commission together with Caspian Littoral states and their neighbours in order to enhance cooperation in energy and transport; and since the 2007 launching of the Black Sea Synergy, the Commission has been working diligently to integrate the Black Sea region into the European energy market. All of these initiatives play to Europe's strengths by aiming at the export of European market principles and regulations (and 'good governance'), which happens to be widely perceived as the EU's most potent negotiating tool in international energy negotiations. However, they can also be seen as accompanying measures to a more proactive energy foreign policy. As the High Representative for the Common Foreign and Security Policy, Javier Solana, said in 2006, Market liberalisation 'is only part of the answer' (Solana 2006: 2).

By 2010, the European Union undoubtedly was one of the world's leading bodies on matters of energy and the environment, much of which was due to decades of tireless efforts by the Commission to promote the needs and benefits of a Community-wide, and now Union-wide, sustainable energy policy. Therefore, when negotiations at the 2009 Copenhagen Summit failed to materialize any binding outcomes, the member states decided to push forward with their own solution. In March 2011, the European Commissioner for Climate Action, Connie Hedegaard, introduced *A Roadmap for Moving to a Competitive Low Carbon Economy in 2050* (the so-called Low Carbon Roadmap) that examined cost-effective ways of reducing greenhouse gas by 80–95% by 2050, compared to 1990 (European Commission 2011a).

Nine months later, responding to a request from the Extraordinary European Council of 4 February 2011, the Commission adopted the Communication *Energy Roadmap 2050* that provided the basis for developing Europe's next long-term energy policy framework, together with stakeholders throughout the energy sector (European Commission 2011b). As expected, the roadmap identified the three familiar goals of comprehensive energy security (decarbonisation, supply security, and competitiveness). However, it went further, explicitly claiming that a secure, competitive, and decarbonised energy system could be possible by 2050 as long as certain conditions were met. Those conditions include fully implementing the EU's 2020 strategy, massively increasing energy efficiency and the share of renewables, increased and sustained public and private investments in research and development, and more coordination in international energy relations. In short, the roadmap provided a shopping list of the Commission's long-standing positions on liberalization, energy, and the environment. It also positioned the Commission to set the agenda of future negotiations in the energy sector.

In October 2014, the member states agreed on another framework for climate and energy policies. Among the usual suspects were the reduction of greenhouse gases, the increase of renewables, and the boosting of energy efficiency (European Council 2014). Yet as is so often the case, the devil can be found in the detail. The preference gap between the member states was so vast that some aspects of the 2030 Framework were assigned to the authority of the European Council, which acts by consensus, such as negotiations on the further development of the ETS, investments into the infrastructure (interconnectors), and mandatory increases in energy efficiency. This is particularly interesting because it is in essence, *contra legem*. All energy provisions should be adopted in accordance with the ordinary legislative procedure (Art. 194 TFEU), that is, the EP and the Council acting upon a bill initiated by the Commission.

Uncertainty abounds about how likely it is that the Commission's long-term predictions will be realized, and while establishing a low-carbon economy seems to be desirable from an environmental point of view, some renewable energy technologies remain simply uncompetitive. Given the urgent need to secure the energy supply and the latest round of retreats from nuclear power by some EU countries, particularly Germany, it may prove problematic to stick to the Commission's 2050 vision.

Concluding remarks

This brief historical overview shows how the creation of a European internal energy market cycled between ambitious Commission proposals, rhetorical commitments by the European Council, and a suboptimal dedication to substantial changes by the Council. In the early years the tensions between those actors were accentuated by the enormous needs associated with Europe's post-war reconstruction. Coal was the energy source of the day. In the 1960s, there was little progress because energy issues simply were not very high on the political agenda. Cheap oil and optimism about the future of nuclear energy rendered the idea of an internal energy market intriguing, but not urgent. The oil crises of the 1970s changed everything. Since then, European energy policy, understood as the sum of all national and supranational energy policies, has been driven by three broad aims: creating a fully integrated and liberalized European energy market; securing stable energy supplies at home and, where necessary, from abroad; and mitigating damage to the environment whenever and wherever economically feasible; in other words, the multidimensional pursuit of comprehensive energy security. There were many significant leaps forward and even more points of retreat in the building of a European energy policy. Its last great leap forward was its entry into primary law in the Lisbon Treaty, and its most recent setback has been the 2014 crisis over Ukraine.

Much of the credit for the progress goes to the Commission, which tirelessly pushed for greater understanding among the member states on the interdependence between energy security, economic stability, environmental conditions, and social welfare. Along the way, the Commission seized the opportunities that presented themselves to promote energy market liberalization, particularly in the gas and electricity sectors, through price transparency, coordinated investments in infrastructure, and the opening of the existing grids to alternative suppliers. It has led the way on the environment and where possible, used environmental laws to achieve its energy agenda. On top of its legislative and market achievements, the Commission has built over the course of more than five decades, an admirable, wide-ranging consultative network consisting of energy stakeholders, from producers, experts, and member state representatives to consumer groups. While prices and regulations still vary according to national needs and energy mixes, the variation is much less than it was decades earlier. Furthermore, European consumers have greater choice over who delivers which type of energy to their homes than at any time in the past. Nevertheless, resistance to the internal energy market project persists.

With respect to integration theory, the historical milestones show the complexity of the European integration process, a complexity too vast to be grasped by one single theoretical approach. One finds elements of Liberal Intergovernmentalism, Functionalism, and Institutionalism at play in the various stages of the energy policymaking process. National positions depend on national actor constellations, and the success of Commission initiatives depends on those national actor constellations, as well as external shocks. The internal dynamics of the European Council are dominated by 'high politics' (i.e. the international context) and national considerations. All of these institutions act within their very own bounded rationalities, resulting in a complex multilevel interplay of different political logics.

Finally, one would be remiss to underestimate the impact that external events have had on the evolution of Europe's energy policy. Instability in the Middle East and North Africa, the growing importance of the Caspian Littoral and the Arctic Sea, increased international competition for oil and gas, and the deeply contentious international climate change negotiations had and will have wide-ranging effects on its pursuit of all three pillars of comprehensive energy security. The creation of an internal energy market, for example, is not only a question of sustainable energy supply at affordable prices but also a question of strategic security in case of energy shortages. The only way to counteract the risk of disruption to the flow of supplies in any part of the Union, or from any one energy corridor, is to have the ability to move energy across the Union unimpeded; and that will require large investments into cross-border infrastructure. Pollution is a negative externality of energy production, no matter what the source or location. Some sources are cleaner and some are cheaper, but neither are both, and location merely delays its effect. Therefore, all energy initiatives, whether internal or external, will need to be executed alongside sustainable environmental policies; and those policies have to make economic sense. Lastly, in terms of supply security, it stands to reason that a Union of 28 member states would be best served if it were to develop a common external energy policy. Such solidarity would maximize its pull as a consumer and minimize the centrifugal tensions that tear at the Union's cohesion, as well as reduce other risks associated with individual bilateral energy relations. However, despite the fact that such 'a spirit of solidarity' is in principle enshrined in primary law (Art. 194 TFEU), 'energy solidarity' among the 28 member states remains very much a work in progress.

Chapter 4

Who Does What? The Main Actors

This chapter examines the actors involved in EU energy policy-making and illustrates the interdependence between the major players in the policy process, including formal and informal actors. Who are these key actors? How do they relate to one another in energy policymaking, and what instruments are available to them? Due to the divergent degrees of Europeanization in the different areas of European energy policy, the potential to influence the decision-making process differs considerably between actors. Energy policies in the EU primarily remain the responsibility of the member states and are an essential element of domestic politics, not least because any domestic economy is dependent upon reasonably priced electricity for manufacturing and private consumers need affordable home heating as well as fuel for their cars. Thus, domestic lobbying groups in modern democracies try to exert as much pressure as legally and reasonably possible on their governments in order to ensure that the actions and policy choices of that government (internal and external) reflect their interests. Governments, in turn, attempt to realize domestic interests in international negotiations and, in so doing, attempt to mediate between different levels. Putnam describes this two-level game from the perspective of government leaders as follows (1988: 434):

> Across the international table sit his foreign counterparts, and at his elbow sit diplomats and other international advisors. Around the domestic table behind him sit party and parliamentary figures, spokespersons for domestic agencies, representatives of key interest groups, and the leader's own political advisors.

Indeed, energy policy touches upon so many adjacent policy fields, from competition policy to industrial policy, environmental sustainability, etc., that the space for supranational policymaking expands considerably. Meanwhile, the development of internal EU energy policies both affects and is affected by the global energy business. Thus, it behoves individual member states to maximize their influence on any matter that carries direct consequences for the external dimension.

Within the institutional arrangement of the EU, the member states have at least two main official opportunities to articulate their interests: the Council (e.g. Transport, Telecommunications and Energy Council (TTE)), also informally known as the 'Council of Ministers', and the European Council assembling the heads of state and governments. Since the Treaty of Lisbon (2007), the ordinary legislative procedure is applied to ever more policy areas, which also gives the European Parliament, representing EU citizens, a strong role in the policy process. Before taking a closer look at the European Council and the Council, and the role they play in EU energy policymaking, however, we must first focus on the European Commission. Not to be underestimated in the entire integration process is the Court of Justice of the European Union, which, tasked with the interpretation of EU law such that it is applied uniformly across the member states, was crucial in developing the EU's internal energy market.

The European Commission

The Commission is a supranational body established for acting independently from the member states in the interest of the entire Union. It is the only institution responsible for initiating EU law and then later ensuring its application by the member states. The Lisbon Treaty strengthened the Commission's executive power, particularly in relation to day-to-day decisions (such as the delegation of powers by the Parliament and Council on a case-by-case basis to adopt quasi-legislative measures). Whenever the Council and the European Parliament pass legislation, it is the Commission's bailiwick to survey whether and to what extent the member states correctly apply EU law.

As the EU's executive branch, the Commission oversees multiple administrative divisions called Directorate-Generals (DGs), which are comparable to national ministerial portfolios. While the DGs work semi-autonomously, their legislative drafts must be agreed upon by the entire College of Commissioners. Frequent and regular meetings between the DGs beforehand facilitate consensus-building at the level of Commissioners. Of greatest relevance to the EU's energy policy are the respective DGs for Energy (ENER), Climate Action (CLIMA), Competition (COMP), Mobility and Transport (MOVE), Enterprise and Industry (ENTR), and Environment (ENV), a few of which are recent creations, established by the Barroso Commission (2010–2014) in 2010. Specifically, the DG for Energy and the DG for Mobility and Transport are the successors of the former DG for Energy and Transport, which prior to 2010 was responsible for all activities related to energy services and mobility. Similarly, the DG for Climate Action took over administration of climate-related topics that were previously the responsibility of the DG for the Environment. Despite the administrative separation, however, policy fields and issues continue to overlap. This is apparent, for example, when one thinks of the importance of fuel prices for logistics companies or the amount of greenhouse gas emissions by the transport sector.

The new European Commission (2014–2019), under the leadership of Jean-Claude Juncker, embarked on a significant restructuring of the Commission's organization by introducing clusters or 'project teams', each headed by one of the vice-presidents. Maroš Šefčovič heads the team on 'A Resilient Energy Union with a Forward Looking Climate Change Policy'. In his inaugural speech, Juncker listed the creation of an Energy Union amongst his top ten priorities (Juncker 2014). Pooling resources, combining infrastructures, and uniting negotiating power vis-à-vis third-party countries are the cornerstones of this new and arguably ambitious undertaking.

Insecurity resulting from Russia's intervention in Ukraine forced the Commission to look beyond the implementation of existing legislation and once again turn its attention to energy security. The Commission ordered so-called gas stress tests in May 2014 (European Commission 2014), which led the Commission to issue a recommendation to the Council the following October (European Commission 2014r). The lessons learned were unambiguous. Without cooperation amongst the member states, a disruption of gas deliveries would potentially have disastrous effects, especially on the Eastern member states. The Commission also made it clear that it prefers a market-based approach. Non-market measures, such as the release of strategic stocks, forced fuels switching, and demand curtailment, should only kick in when the market fails.

Consequently, one of the most important actors is the DG for Competition, which plays a significant role in the internal market project. EU law authorizes the Commission to impose fines in case of violation against competition law (Art. 101ff. TFEU), giving the Commission considerable competences in this area. It is not surprising, therefore, that it continuously tries to strengthen the EU's role in energy policy through competition policy. For example, in 2009, the Commissioners imposed fines totalling €1,106,000,000 on three companies, Germany's E.ON AG and its subsidiary E.ON Ruhrgas AG, and on France's GDF Suez SA, arguing:

> Ruhrgas AG (now E.ON Ruhrgas, part of the E.ON group) and Gaz de France (now part of GDF Suez) agreed in 1975, when they decided to jointly build the MEGAL pipeline across Germany to import Russian gas into Germany and France, not to sell gas transported over this pipeline in each other's home markets. They maintained the market-sharing agreement after European gas markets were liberalised, and only abandoned it definitely in 2005. These are the first Commission fines imposed for an antitrust infringement in the energy sector. (European Commission 2009)

In addition to its competences in competition policy, the Commission possesses further instruments to enforce compliance. For instance, according to Art. 258 TFEU, the Commission has the duty to ensure that the member states correctly apply EU law. In case a member state fails to comply with EU law, the Commission has the right to take appropriate measures to make the state fulfil its obligations. In the pre-litigation phase, known as 'infringement proceedings', the Commission prepares a letter of formal notice in which it requests that the member state explain the reasons for non-compliance, providing the opportunity to comply voluntarily within a given time frame. If member states continue to ignore their obligations under the treaties, the Commission has the discretionary power to refer a case to Court. However, the mere threat of infringement proceedings often serves to improve a member state's performance in applying EU law. As the Commission made clear in its first attempt to narrow down the concept of the Energy Union, it considers the strict implementation of the Third Energy Package the cornerstone of its endeavours.

The European Council

Alongside the increased influence of the Parliament and its committees and the long-standing influence of the Council and its working parties, the role of the European Council has been strengthened in recent years, particularly in energy and climate change matters (Christoffersen 2009: 84). The European Council sets the EU's overall political directions and usually makes decisions unanimously. Therefore, it can be assumed that all member states support the adopted measures, which facilitates implementation at the national level. Nevertheless, the decisions reached are often based on a minimum consensus, because member states' interests diverge substantially with regard to EU energy policy in general, and particularly as far as the function of the European Union is concerned. For example, the member states agreed on the necessity to unbundle vertically integrated energy companies. However, concrete measures required to realize this objective continue to be subject to highly controversial discussions (Pollak et al. 2010).

The European Council's importance has risen dramatically over recent years in sync with EU–internal crises, such as the hard-hitting recession since 2008 and international crises ranging from the 2003 Iraq War to the annexation of the Crimea by the Russian Federation in 2014. However, while the EU's energy interests were clearly affected by all those crises, its policy responses were geared more towards general statements and declarations. Those responses mirror the divergent preferences of the member states. Given the contentious nature of external dimension issues, such as bilateral supply contracts, that directly influence the different security supply interests of the member states (i.e. the old problem of who supplies whom under which contractual terms), a unified position would be short of a miracle.

In April 2014, Donald Tusk, then prime minister of Poland, proposed an Energy Union in an article in the Financial Times (Tusk 2014). At the end of 2014, Tusk assumed the office of President of the European Council and is certainly one of the driving forces behind the idea of an Energy Union. Meeting in March 2015, the European Council came up with a vague agreement to establish an Energy Union (see European Council 2015), which once again committed the EU to build an affordable, secure, and sustainable energy market within the EU. Shortly thereafter, the Commission sent a Statement of Objections to Gazprom alleging that some of its business practices in Central and Eastern European gas markets were in breach of EU antitrust rules, because they constituted an abuse of its dominant market position. Thus, it can be surmised that the European Council and the Commission share the same strategy: continue the integration of the European energy market, increase cross-border trade, and secure the compliance of Europe's energy acquis for all players.

The Council

Within the institutional arrangement of the EU, the member states also have the opportunity to articulate their interest in the Council. The Council has a huge impact on the policy process, because it is the institution where the representatives of the member states meet regularly, and this is why it has been described as the EU's real 'decision-making centre' (Wessels 1991). The Council, in its ten different configurations, consists of the relevant ministers of the member states, who discuss the subjects that are on the EU's agenda. The frequency of the Council's meetings depends on the configuration (see Council configurations at http://www.consilium.europa.eu). While the ministers in the Foreign Affairs Council, dealing with the EU's external action, or the Ecofin Council, responsible for all economic and financial affairs, usually meet once a month, other issues require less frequency.

Energy issues are dealt with primarily in the Transport, Telecommunications and Energy Council (TTE), established in 2002, and meets approximately every two months. Consisting in its different configurations of the relevant member state ministers, it debates and, indeed, represents the EU's main goals in transport, telecommunications, and energy policy. These goals are explicitly defined as establishing 'modern, competitive and efficient markets and infrastructure, and to create trans-European transport, communications and energy networks' (Council 2015), which should contribute 'to the smooth operation of the internal market and to strengthening economic and social cohesion' (EP/Council 2010c). The TTE's bailiwick places it amidst all three dimensions of the energy policy. As far as its energy policy component is concerned, the Council promotes five goals: energy supply security for all EU consumers at affordable prices, energy efficiency, fair competition in the internal market, the peaceful use of nuclear energy, and environmental protection. One example of this is a recent directive adopted by the

Council that concerned environmental protection. This directive on the safety of offshore oil and gas operations (2013/30/EU) aimed at minimizing the consequences of accidents on the marine and coastal environment (EP/Council 2013). It lists safety standards' criteria for offshore oil and gas drilling, particularly in the light of the May 2010 Deepwater Horizon oil spill in the Gulf of Mexico, and sets rules whereby if offshore producers fail to apply the standards, no further licences will be issued, or worse, the company in question will face penalties ultimately resulting in the loss of their existing licences. The TTE adopted the act on 6 June 2013 after the European Parliament's first reading.

Before the TTE either discusses or decides upon matters on its agenda, proposals are run through preparatory bodies known as working parties (sometimes referred to as working groups). These committees 'may be set up, or with the approval of Coreper, with a view to carrying out certain preparatory work or studies defined in advance' (Council 2009b: 13).

Directive 2009/28/EC (Article 17(2)–(6)) set the criteria for biofuels (and bioliquids) in order to ensure sustainable production (European Commission 2012b). The sustainability criteria identified by the TTE were quite specific about emissions savings from the use of biofuels and bioliquids (35% to start with and 60% by January 2018), as well as putting a curb on what type of land could be used for biofuel crops; specifically noting that such fuels 'shall not be made from raw material obtained from land with high biodiversity value' (EP/Council 2009d). The European Council's Atomic Questions Group (AQG) set up another committee, the Working Party on Nuclear Safety (WPNS), at the end of 2000 to evaluate nuclear safety standards in candidate countries in the context of enlargement. That Committee made a 'technical evaluation of the information made available to date (1/05/02)' and found that the 'Candidates Countries are clearly committed to fulfil the recommendations set out in a report on nuclear safety (Council 2001), both for NPP and other types of installations' (Council 2002: 5). The cases above are indicative examples of how important the Council's committees are in terms of defining the technical details, targets, and wording of key energy legislation, a role that should not be underestimated in such a complex and diverse policy field.

Three other working parties bear mention here due to their noticeable influence or nominal ranking, if not lack of transparency. There are two important working parties responsible for financial outlays: the Budget Committee, which approves financial allocations together with the EP, and the Working Party on Tax Questions, which decides on energy-related taxation. As influential as both are with regard to the traditional power of the purse and the always thorny tax question, there is one working party that can easily be considered the TTE's most important for energy matters: the High-level Working Party on Energy, which also appears under the simple name Working Party on Energy, depending logically on its configuration (Council 2014). Widely considered to be one of the most influential working groups, it is also one of the least documented energy specific, because its meetings are generally off the public's radar. It consists of national experts from related industries in the 28 member states, who represent their interests and prepare technical dossiers to the ministerial level, which then holds exclusive authority to make legally binding decisions. In early 2015, the Council embarked on a discussion of priorities for Europe's energy infrastructure, a move that was in line with the European Council's agreement on setting up an Energy Union and the Commission's recommendations after the gas stress tests.

The European Parliament

The European Parliament's relevance in the decision-making process increased substantially following the 1993 introduction of the co-decision procedure (see Maastricht Treaty). The Lisbon Treaty subsequently solidified its role as a primary actor in EU energy policy by providing an upgraded version of the procedure, known as the *ordinary legislative procedure*, which gives the EP the same weight as the Council in the policy process. These changes have had a real impact on EU energy policy.

Most of the EU's daily work on energy policy is done in the Committees of the European Parliament (a complete list can be found on its website, www.europarl.europa.eu). The most decisive ones are the Committee on Foreign Affairs, the Committee on Industry, Research and Energy, and the Committee on Environment, Public Health and Food Safety, all of which are Standing Committees. Where necessary, committees must resolve differences before legislation can move forward, a representative feature of the EP's expanded role in EU energy policymaking. For example, in order to pass the legislation on measures to reduce CO_2 emissions from cars to a binding target of an average 130 g CO_2/km by 2012, as proposed by the Commission in December 2007 (European Commission 2007k), the Environment and Industry Committees had to overcome contrasting views before the European Parliament could agree on a common position. The debate was an old one. Passenger car emissions had been a subject of EU concern since 1995, when they were emitting 186 g CO_2/km on average (European Commission 1995b). While the Industry Committee opted for less strict regulations for emission performance standards, the Environment Committee favoured a 120 g CO_2/km target for passenger cars (ten Brink 2010). Ultimately, the European Parliament approved the text of the legislation, but the debate took almost a year (EP/Council 2009e).

Another case occurred in January 2014, when MEPs gave mixed responses to the Commission's proposals for measures on how to tackle climate change. Although the MEPs of the Environment and the Industry Committees in general 'backed a 40% cut in CO_2 emissions, a 30% target for renewable energy and a 40% target for energy efficiency for the period to 2030' (European Parliament 2014a), individual opinions on the effects of the proposals diverged. For example, the Chairman of the Environment Committee, Matthias Groote, said, 'Apart from the 40% greenhouse gas reduction goal, these targets still fall short of what is needed [...]' and the co-rapporteur for the Industry Committee, Konrad Szymanski, critically claimed that 'increasing the binding target for energy from renewables to 27% does not take into account the electricity price impact of this policy' (ibid: 1). Such lively debate at the Committee level is a common feature of energy policymaking in the EU, one that is indicative of the EP's growing importance in both formulating policy as well as providing the basis for sceptical oversight.

In addition to these standing committees, the EP on occasion sets up temporary committees. In March 2007, for example, it set up the Temporary Committee on Climate Change (TCCC) to develop scenarios on how to design the EU's future climate change policy and demonstrate the EP's seriousness about tackling the challenges imposed by global warming (European Parliament 2008). The TCCC initially had a one-year mandate, but in February 2008, its mandate was extended for another nine months.

Another type of relevant EP body is the parliamentary intergroup (PI), which constitutes an important, albeit informal, forum for exchanges between parliamentarians and civil society. The Conference of Presidents set the rules governing the establishment of intergroups in 1999 (European Parliament 1999). Such bodies complement the work of committees. For example, the European Parliament Intergroup on Climate Change, Biodiversity and Sustainable Development provides a forum for the MEPs to discuss current energy policy and climate change developments and jointly develop strategies. Different political groups support the intergroups, which include members from all political colours and parliamentary committees. As parliamentary intergroups are not formal parliamentary bodies, positions issued by an intergroup do not express the opinion of the European Parliament. Indeed, according to the Rules of Procedure for the 7th parliamentary term (2009–2014), 'such groupings may not engage in any activities which might result in confusion with the official activities of Parliament or of its bodies' (European Parliament 2014: 28, Rule 32). For a complete list of current intergroups, see the EP's website (www.europarl.europa.eu).

Court of Justice of the European Union

At the EU level, the Court of Justice of the European Union (CJEU) is another decisive player in EU energy policy, particularly with regard to the establishment of the internal energy market. For example, in 1994, the CJEU defined electricity as a 'good within the meaning of Article 30 of the Treaty' (C-393/92) and, in so doing, enabled its integration into the internal market regime. The same applied to gas.

More recent cases concerning the internal market deal with Bulgaria's violation of the gas transmission regulation (C-198/12), action brought by the Commission against Poland over the prohibited regulation of gas prices (C-36/14), as well as preliminary rulings over unilateral gas price adjustments in Germany (C-92/11; see also C-359/11 and C-400/11), and the unbundling of gas distribution system operators in the Netherlands (joint cases C-105/12 to C-107/12). The Court also delivered important controversial rulings concerning renewables, such as Alands Vindkraft vs. Energimyndigheten (C-573/12), which confirmed the right of the member states to subsidize renewable energy sources in accordance with national regulations. With this ruling, the Court allows 28 different subsidy schemes for renewables, which is in contention with the objective of a EU-wide integrated electricity market. However, current EU law justifies this restriction in the application of the internal market rules because the EU has not yet harmonized national support schemes for green electricity, which results, for example, in green electricity producers in Germany being unable to receive subsidies for electricity they sell in France. This restricts one of the four fundamental freedoms of European integration, namely the free movement of goods. Other cases concern the energy performance of buildings, Directive 2002/91/EC (European Parliament and Council of the European Union 2002), (e.g. C-329/14 and C-302/14), energy labelling, Directive 2010/30/EU (e.g. C-319/13 and C-609/12), and the international application of the ETS (see Chapter 6, Section "Case study: The EU emissions trading regime').

The CJEU is also involved in the interface between the state and the market (Schmidt 2003) where it deals with the question of the extent to which the member states are allowed to confer privileges to companies that provide services of general interest and, in this way, act beyond the limits of competition law (Weiß 2003). Contrasting views exist on this point because the guidelines in the TFEU (Article 106, Paragraph 2) are rather vaguely formulated. This latter point is of special importance for the Commission, which seeks to minimize the regulatory competences of the member states while the member states tend to apply a broad interpretation of EU law.

One might think that the CJEU and Commission are natural allies. However, while the CJEU initially supported the Commission on many issues, the Court is increasingly granting the member states more rights when it comes to the provision of common goods, particularly since the 1990s. This tendency is reflected in its PreussenElektra judgment of 13 March 2001 (C-379/98) concerning the compatibility of the German support scheme for national renewable energy. However, the Court issued a somewhat contradictory decision a little over a week later (21 March 2013) on a case concerning the transparency of gas price increases for consumers. In this case, the CJEU decided that it was up to the court at the national level to determine in each individual case whether the contractual terms used by a company (Germany's RWE in this case) comply with the generally acknowledged requirements of good faith, balance, and transparency. The consumer association for North Rhine-Westphalia (*Verbraucherzentrale Nordrhein-Westfalen*) challenged the use of an allegedly unfair term in contracts with special consumers, by which the RWE reserved the right to unilaterally adjust gas prices. Based on its contractual terms, RWE increased those prices on four occasions between July 2003 and October 2005. In its decision, the Court found 'that the EU legislature recognised that, in the context of contracts of indefinite duration such as contracts for the supply of gas, the supplying undertaking has a legitimate interest in altering the charges for its service' and 'points out that ultimately it is not for it but for the national court to determine in each individual case whether that is so'. (Court of Justice of the European Union 2013: 2).

Transnational NGOs, lobby groups, and informal fora

Another body of actors play a less formal, but no less important, role in developing and implementing EU energy policy. These include transnational non-governmental actors (TNGOs), peak environmental and economic associations (and lobby groups), and informal fora, each of which provide a bottom-up means for EU stakeholders to frame and influence policy debates and outcomes.

Transnational (often EU-wide) non-governmental actors influence the policy process not only after the Commission presents a proposal for further discussion but also, and most importantly, during the implementation phase. Here one has to differentiate between the largest and best financed groups, such as the large environmental and economic associations and those that are rather fragmented and less equipped with financial and human resources. Furthermore, many well-resourced, multinational companies have established their own office in Brussels, which is often done concurrent to membership in various associations and, thus, allows direct participation in the daily activities of EU policymaking, ensuring that their specific views are taken into account in an efficient and timely manner.

Interest groups can pursue different strategies to influence decision-makers depending on the amount of capital they can invest in lobbying. For example, large companies are usually members of sectoral, national, and European associations, which serves to fulfil two distinct functions. First, associations educate their members by monitoring the (legislative) developments at the EU level and communicating their findings to their members. Second, associations lobby policymakers in the interests of their constituencies by submitting positions summarizing their members' common views and/or articulating their positions in personal meetings with the representatives of the EU institutions. Highlighting common interests, giving recommendations for further action, and following a long line of well-practised traditions, they are often successful.

The biofuels sector provides a good example of non-governmental influence in EU energy policy. Associations and companies representing the industry's interests were very active in lobbying the Commission's October 2012 draft law on the future of first-generation biofuels (European Commission 2012b). The fuels involved were often criticized for being pollution-intensive and for displacing food crops in developing countries, including sugar cane and corn ethanol, and oilseed rape and palm oil biodiesel, among others. This problem is also known as indirect land-use change (ILUC), which can be understood as the net carbon loss from the clearance of forests and grasslands for necessary new food production as a result of biofuel cultivation. Already in 2008, the International Energy Agency (IEA) raised concerns regarding the implications of the production of these biofuels for the environment and malnutrition in developing countries. Accordingly, ethanol and biodiesel production were assumed to have negative impacts on food security and food prices, scarce water resources, deforestation, and biodiversity (IEA 2008: 6). In an attempt to overcome the opposition against them, the biofuels industry at one point sent 'three mails an hour, many containing catastrophic warnings' in order to make sure that the Commission took their interests into account (Euractiv 2013). While under the email barrage, the Commission began considering the production of second-generation biofuels, thinking that their development would be more promising, if not less problematic, even though it is generally considered unlikely that second-generation biofuels will be produced commercially before 2015 (e.g. those derived from cellulosic materials such as biomass to liquid (BtL) and cellulosic ethanol (IEA 2008: 33)). In the end, the EU imposed a 5% cap on the amount of first-generation biofuels in the EU's 2020 transport mix and imposed a 60% greenhouse gas-saving threshold starting in July 2014, but did not include feedstock-specific values – a small, but important victory for the industry.

In order to make the European decision-making process more transparent with regard to the actors who are influencing the EU institutions and the particular interests they pursue, various attempts have been made to set up lobbying registers. In 2001, the Commission issued a White Paper on European governance that emphasized a number of core principles as essential for more democratic governance. In it, the Commission wrote:

> Roles in the legislative and executive processes need to be clearer. Each of the EU Institutions must explain and take responsibility for what it does in Europe. But there is also a need for greater clarity and responsibility from Member States and all those involved in developing and implementing EU policy at whatever level. (European Commission 2001b: 10)

The European Parliament and the European Commission established a transparency register in 2011 that lists organizations and self-employed individuals engaged in EU policymaking and policy implementation (EP/European Commission 2011). In May 2015, it listed 7,295 registrants (EP/European Commission 2015). By registering, organizations and individuals indicate that they are interested in contributing to the EU decision-making process and, thereby, follow a certain code of conduct (EP/European Commission 2014).

Companies and associations often coordinate their action with others and form coalitions in order to benefit from synergy effects. In so doing, they also follow the advice of the European Parliament to 'Work with others – look for allies!' This recommendation was clearly expressed in a document published on lobbying tactics at the EU level with the argument that 'officials, parliamentarians and Members of the Economic and Social Committee and Committee of the Regions get very frustrated when more than one pressure group says more or less the same thing but in different words', efforts that lead 'to horrible confusion' (European Parliament 2003: 22).

When actors do decide to coordinate their activities and cooperate, these coalitions can be of an informal, semiformal or formal nature (Kreutler 2014). Informal coalitions are usually based on common short-term interests and aim at influencing the political decision on a specific policy issue. For example, European associations representing the aviation industry created several informal ad-hoc coalitions in response to attempts to include aviation in the EU-ETS. Importantly, it is often the case that such coalitions, which usually publish joint position papers or reports, neither have a name or letterhead, nor a spokesperson or an administrative body to coordinate the work between their members.

Semiformal coalitions form to influence multiple political decisions on different topics. In contrast to informal arrangements, such a coalition has a name that emphasizes its members' common concerns, but no common administrative body. An example of this is the Alliance of Energy Intensive Industries – the name, of course, implying a certain degree of formality. Nevertheless, this particular coalition neither set up a website nor designed a specific letterhead, lacked an administrative body to coordinate the work between its members, and did not identify a spokesperson that is allowed to speak on its behalf. Indeed, the Alliance of Energy Intensive Industries cannot even be found in the aforementioned transparency register. Working on an 'apparent-as-needed' basis, the coalition seems to form in various constellations at various points in time, based on the policy under discussion, such as during the political negotiations about the EU-ETS or the creation of an internal energy market (European Alliance of Energy Intensive Industries 2004, 2006, 2007).

In contrast to their informal and semiformal siblings, formal coalitions are characterized by their formal structures, most notably an established administrative body to coordinate the joint efforts of its members. Formal coalitions are not merely based on specific short- or long-term interests. Its members frequently share and support a common set of norms and thus collaborate on a broader norm-based agenda. A good example is Green 10, a formal coalition of ten large European environmental NGOs including, among others, Greenpeace and International Friends of Nature (IFN), that has actively and successfully cooperated over years on issues as diverse as the inclusion of the aviation industry in the EU-ETS (Green 10 2005: 1), emissions from light duty vehicles (Green 10 2007), and the European Commission's general performance in meeting environmental challenges (Green 10 2009). Logically, such coalitions usually have a profile in the transparency register.

National regulatory authorities

At the national level, national regulatory authorities constitute another group of important actors in EU energy policy. National regulatory authorities (NRAs) are public bodies that are independent of both industry and government, have statutory authority under EU law, and operate in every member state. As NRAs are responsible for implementing rules for their respective energy markets in line with EU regulations, this particular group of actors plays an essential role with regard to the EU's internal energy market.

EU law obliges the member states to establish independent national regulatory bodies (NRAs) in order to ensure fair competition within the market (Pollak et al. 2010). Early rules on NRAs set out in the first two legislative packages of 1996/98 and 2003 were quite vague concerning structure and responsibilities and, thus, there were large differences between their competences and levels of cooperation. The third legislative package of 2009, though, corrected these problems and strengthened cooperation between them. The Commission had established the European Regulators' Group for Electricity and Gas (ERGEG) in 2003 to advise it on internal energy market issues (European Commission 2003). This was then succeeded by the Agency for the Cooperation of Energy Regulators (ACER) in July 2011, as part of the operationalization of the third legislative package (2009), in order to further coordinate and intensify the dialogue between the national bodies and overcome divergent positions between the stakeholders, which impede the integration of the markets. ACER emphasizes some of these challenges vis-à-vis the internal energy market by noting:

> The EU aims to fully integrate national energy markets by 2014, to give consumers and businesses more and better products and services, more competition, and more secure supplies. Progress has already been made: consumers can switch suppliers for gas and electricity, and suppliers must provide clear explanations of terms and conditions. Work still to be done includes aligning national market and network operation rules for gas and electricity as well as making cross-border investment in energy infrastructure easier. (ACER 2014)

Like the ERGEG before it, ACER advises the Commission on internal energy market issues, but has permanent staff and experts seconded by national regulatory authorities for energy, which was something the ERGEG lacked. Similar to its predecessor, however, it has a Board of Regulators, consisting of senior representatives of each NRA plus one non-voting representative of the European Commission. Observers regularly join their meetings, including the heads of NRAs from candidate countries as well as those from countries belonging to the European Economic Area (EEA).

The structure and functions of ACER overlap with those of the Council of European Energy Regulators (CEER), and the two share similar objectives. However, ACER is a formal EU agency funded by the EU, whereas the CEER is organized on the basis of voluntary cooperation between national regulatory authorities. The latter, which positions itself as the voice of Europe's national regulators of electricity and gas at EU and international level, aims to facilitate 'the creation of a single, competitive, efficient and sustainable EU internal energy market that works in the public interest' (CEER 2014).

Two additional, although less formal, fora for information exchange require mentioning. The first is the Electricity Regulatory Forum (also known as the Florence Forum) and the second is the Madrid Forum on Gas. The former was established in 1998 to facilitate the discussions on the internal electricity market. Its participants range from member-state governments, national regulatory authorities, and representatives from the Commission, to TSOs, electricity traders, consumers, network users, and power exchanges (Florence Forum 2015). As of 2013, the forum was focusing on questions about the cross-border trade of electricity. The Gas Regulatory Forum (Madrid Forum) was initiated in 1999 to serve as a platform for information exchange concerning the gas market. Similar to the developments in the Florence Forum, the Madrid Forum was addressing the cross-border trade of gas in late 2013. In its conclusions of October 2013, the forum emphasized the critical role of gas in the light of current decarbonization tendencies and recognized 'the difficulties to deliver the expected benefits in current market conditions and inter-fuel competition dynamics' (Madrid Forum 2013: 1).

Case study: The EU's eco-design directive

The preceding discussion provided an overview of the main actors and actor-types active within the EU's energy policymaking process. Member-state governments directly affect policy through the Council, in its various configurations, and the European Council. European citizens are represented through the European Parliament. Embodying the collective interests of the Union, the Commission is the sole initiator of any legislation, and the Court of Justice of the European Union, through its decisions and rulings, acts to ensure the integrity of the Union's legal principles. On the surface, most EU policymaking appears to be an output of their interaction. Yet, much more is happening beneath the surface; not only in terms of the roles played by industry associations and other special interest groups but also within and between the Council configurations and Directorate-Generals of the Commission.

Understood in all its complexity, the interaction within and between the EU's core policymaking institutions and other interest-focused players at the national and European levels is just as important in shaping policy outcomes as the ordinary legislative procedure. One good example of this process is the almost two-decade long pursuit and refinement of the EU's eco-design directive. That directive aimed at reducing the environmental impact and energy consumption of energy-using products, including electrical appliances, throughout their entire life cycle (EP/Council 2005a) and later all energy-related products (EP/Council 2009j).

Examining the EU's eco-design directive in general, and its original focus on light bulbs in particular, provides a good example of the complex set of actors and interactions involved in the EU's energy policymaking process. Examining it also allows us to see how simplistic notions about energy savings can drive policies and cause policymakers to expend a great deal of political capital on matters that have a lesser impact than expected. Indeed, the Commission's expectations concerning the potential energy savings due to the new light bulb regime may prove to be overly optimistic. Past experience shows that higher energy efficiency does not necessarily lead to energy savings, but rather tends to increase consumption, since higher efficiency impacts the price inversely (i.e. the more efficient an appliance or process, the cheaper it is to use). As we tend to use more of a good or service the cheaper it is, we can observe the paradox effect of falling energy prices leading to more energy consumption, not less – a fact first discovered by W.S. Jevons (2008: 75) in the nineteenth century. This has proven to be a recurring problem in the realm of oil consumption, as increased fuel efficiency standards have offset higher oil prices, leaving overall fuel consumption stable. Nevertheless, lighting accounts for circa 20% of global electrical energy generation (European Commission 2006b), and according to scientific studies referred to in the EU's energy efficiency action plan, high-efficiency LED technology 'could by 2015 save 30% of today's consumption for general lighting and 50% by 2025' (European Commission 2006b). It is no mystery then that EU politicians are attracted by the idea that one could enhance all three pillars of comprehensive energy security by reducing consumption through increased efficiency of electrical appliances.

By the early 2000s, years of economic growth noticeably raised the living standards of millions of Europeans, and the concomitant increase in the use of electrical appliances was taking up a growing share of overall energy consumption. Recognizing the growing problem and seeing an opportunity to reel in runaway growth in electricity demand, the Commission moved beyond the well-known labelling (and grading) of electrical appliances, originally set up in the 1990s, and set its sights on increasing overall energy by removing the least efficient products from the market, focusing specifically on lamps.

The resulting eco-design directive of 2005 was controversial before and after its passing. Due to the diverse set of interests that were affected, EU policymakers compromised, allowing self-imposed measures by the respective industries, which in turn sparked harsh criticism from various stakeholders, one of whom called it 'an invitation to cheat' (Euractiv 2005). Faced with implementation problems, the Commission added two additional regulations in early 2009, establishing a gradual timetable for the phaseout of specific light bulbs (e.g. conventional incandescent and halogen bulbs) from the market by late 2012 (European Commission 2009, 2009b-c), later postponed to September 2018 (European Commission 2015). Since then, the Commission has passed no less than 13 different directives to implement the eco-design directive, covering a wide range of products from dishwashers to water heaters and dryers, and in 2010, the EU passed an updated directive (2010/30/EC) on energy labelling and its implementation measures (EP/Council 2010).

The number of actors involved in formulating the original directive is indicative of the complex and integrative nature of the process. Within the Commission, both the DG Energy and DG Enterprise and Industry played important roles because each oversaw a different set of related products. The initial thrust in the 1990s on energy efficiency in private households, industry, and the service sectors came from DG Energy. The formal proposal for a directive to establish eco-design requirements for energy-using products, particularly lamps, came from DG Enterprise and Industry (European Commission 2003a). Two Council configurations, in the form of Transport Telecommunications and Energy on one hand and Competitiveness on the other, also played key roles. The common position it adopted in 2004 set the basis for the EP's response.

Within the European Parliament, the Associated Committee for Environment, Public Health, and Food Safety (ALDE) played a vital role, leading the way as MEPs called for much stronger measures than proposed by the Commission and approved by the Council. It identified the action required on implementation, design requirements, and market surveillance, adopting 78 amendments in the first reading, a third of which the Commission accepted. During the second reading, ALDE called for the reintroduction of some of those amendments that where originally rejected, a call supported by the Environment Committee (European Parliament 2005). In the end, however, they had to capitulate on many accounts. For a complete account of the legislative actions leading up to the eco-design directive, see the procedure file of the European Parliament's Legislative Observatory (European Parliament 2005a).

All three EU institutions consulted related NGOs, lobby groups, and informal fora across the entire process and received strong responses in favour or against the legislation by different constituencies. NGOs such as the European Environmental Bureau, the World Wide Fund for Nature (WWF), and Greenpeace raised serious objections to the voluntary nature of the original proposal. Conversely, European industry associations such as the European Lamp Companies Federation and the Federation of National Manufacturers Associations for Luminaires and Electrotechnical Components for Luminaires lauded its flexibility (Euractiv 2005).

The passing and implementing of the eco-design directive was one rife with competing interests that struggled to find a balance between competition and the environment. Its case reveals the complexity and depth of the EU's energy policymaking process. Unfortunately, the jury is still out on its effectiveness. Despite the new regulations, annual household electricity consumption across the EU28 actually increased between 2005 (the year that the eco-design directive was passed) and 2012, although this was noticeably less than the 32% increase observed between 1990 and 2005 (Eurostat 2014o) and the 10% increase that occurred between 2002 and 2012 (Eurostat 2014p). In early 2014, the Commission was still consulting a wide and expanding range of stakeholders about the application of the directive.

Concluding remarks

This chapter explains how various interdependent actors are involved in the development of all three dimensions of EU energy policy. We can identify a series of EU institutional actors, including supranational institutions such as the Commission, the European Parliament, the Council of the European Union, and the Court of Justice of the European Union (CJEU). We can also identify permanent and temporary preparatory bodies, such as working parties and parliamentary intergoups, and important informal, semiformal, and formal transnational coalitions of non-governmental actors, as well as formal and informal fora and networks. As demonstrated in the discussion of the EU's eco-design directive, beyond the default roster of institutional actors, a changing subset of others, from NGOs and companies to groups of politicians, participate with varying degrees of success depending on the subject matter.

The impact of any of these actors on the outcome of the policy process depends on the level and frequency of their access to decision-makers. Indeed, access is a precondition for influence. Once access is gained, successfully influencing policymakers depends on the extent to which those decision-makers are receptive to specific positions and ideas. Moreover, receptiveness by policymakers is often limited because the EU's institutions do not have the time or human resources to adopt them all. Non-governmental actors in the EU are aware of this and, consequently, place a great deal of emphasis on finding the right venues (i.e. the most receptive) to pursue their interests (Baumgartner and Jones 1991, Princen and Kerremans 2008). Still, an actor's access to these venues does not necessarily equate to its ability to change a policy outcome according to its preferences. Decision-making remains a continually flowing process. As such, it is difficult to say at which point in time, and for what reason, policymakers have taken into account certain concerns or changed their minds, even for decision-makers.

Chapter 5

Building a Common Internal Energy Market

Installing a common internal energy market is of paramount importance to the EU and its member states, and has been for decades. However, as of today the respective national energy mixes and market prices still vary widely across the Union (see Chapter 2, Section 'The three end uses of energy'). According to the European Commission, a common market would facilitate lower consumer prices, spark investments in vital infrastructure projects – which in turn would allow for easy energy transportation between the member states (and thus mitigate exogenous supply shocks by allowing multiple points of entry by diverse suppliers) – and foster the development of affordable substitutes and renewable forms of energy. In essence, the Commission presents the common energy market as a major piece of the puzzle to cure the ills of Europe's energy malaise; one that consists of high external dependence, the lack of interconnectivity between a regionalized infrastructure, high degrees of state involvement in energy companies, substantial differences between energy prices in the member states, and widely divergent interpretations of how to secure supply. Thus, the internal energy market carries significant implications for both of the other energy policy dimensions (external and multidimensional), and constitutes, through fair competition, a core pillar of comprehensive energy security. What is the real potential of such an internal market? How much can it contribute to achieving the objectives of competiveness, sustainability, and stable supply?

When looking at Europe's diverse national energy markets, one cannot miss their heterogeneity, from the British liberalized version to the rigidly and vertically integrated energy companies that dominate the markets in France and Germany (i.e. the so-called 'national champions'). This chapter examines the Commission's efforts to enhance competition, ensure price transparency, and secure fair prices across the Union, by attempting to break up national monopolies, and enabling and controlling grid access for third parties, through a network of national regulators and European institutions such as the Agency for the Cooperation of Energy Regulators (ACER). We place special focus on the EU's third legislative package and its relevance for creating a single European energy market.

The creation of the internal energy market

The liberalization of the electricity and gas markets has constituted a core element of the EU's agenda since the late 1980s. The Council argued in favour of the establishment of an internal energy market for the first time in September 1986 in a Council Resolution (86/C 241/01) of 16 September 1986 'concerning new Community energy policy objectives for 1995 and convergence of the policies of the Member States', in which the Council formulated guidelines for the energy policies of the member states. These, however, reflected a minimum consensus, as the member states could only agree on a convergence of national policies, not on communitarization (i.e. replacing national laws with Community law). The White Paper on the internal market and the Single European Act of 1987 did not explicitly refer to energy policy. Nevertheless, they apparently influenced the EU's internal energy market project by deepening market integration. In 1988, the European Commission suggested concrete measures for the liberalization of the electricity and gas markets. These were reflected in a directive aiming at improving the transparency of gas and electricity prices charged to industrial end users (Council of the European Communities 1990a), which was regarded as a prerequisite for fair competition. In addition, the Council adopted two directives on the transit of electricity (1990c) and natural gas (1991) through grids, the latter already being subject to highly controversial discussions. Other proposals for directives, in which the Commission called for greater openness, were rejected by almost all member states, except for the United Kingdom, where the liberalization of the national energy market was already part of Margaret Thatcher's deregulation programme (Andersen 2000, Schmidt 1998).

In February 1997 and August 1998, two directives concerned with common rules for the internal electricity and natural gas markets (EP/Council 1996, 1998) came into force initiating the gradual liberalization in both sectors. As both directives were based on the same legislative acts (Council of the European Communities 1990c (for electricity); 1991 (for gas) and in general followed the same intention, their wording was partly identical. Choosing the path of least resistance, the Commission focused first on electricity, achieved the necessary consensus (leading to Directive 96/92/EC) and thus paved the way for further negotiations in the gas sector (Andersen 2000). Both the gas and electricity directives aimed at increasing 'efficiency in the production, transmission and distribution' while at the same time 'reinforcing security of supply and the competitiveness of the European economy and respecting environmental protection' (EP/Council 1996: 1). The legislation, which was based on two previous acts (EP/Council 1996, 1998), focused on grid interoperability (1996: Art. 7; 1998: Art. 5), unbundling and transparency of accounting (1996: Art. 13–15; 1998: Art. 12–13), third-party access (1996: Art. 16–18; 1998: Art. 14–16), and the establishment of an independent conciliation board (1996: Art. 20; 1998: Art. 23).

Although both directives constituted an important step towards the creation of an internal energy market, the Commission was not able to realize many of its original intentions. While the directives stipulated concrete goals to be achieved during the years to come, it also left room for interpretation at the member-state level, a concession to the member states that had the side effect of slowing down the establishment of an internal market (Pollak et al. 2010: 117). To accelerate the process, the Commission decided in 1998 to establish the Florence Electricity Regulatory Forum (Florence Forum 2015) and the Gas Regulatory Forum of Madrid (Madrid Forum 2013), to enhance consensus-building between the various actors in energy policymaking. They usually meet up to twice a year to discuss issues related to the creation of the internal market, bringing the Commission together with regulatory authorities, member-state governments, TSOs, electricity or gas traders, consumers, network users, and power exchanges. During its meeting in Lisbon in March 2000, the European Council decided to push through a second legislative package in order to speed up the liberalization process, resulting in two new directives that included common rules for the internal markets in electricity and natural gas (EP/Council 2003b, c). Both acts repealed their respective predecessors and marked a major step forward in the EU's efforts to integrate its diverse energy markets.

One of the key issues of the new directives was the unbundling of accounts by electricity and natural gas firms (EP/Council 2003b:

Art. 19; 2003c: Art. 17). Accordingly, the companies had to 'keep separate accounts for each of their transmission and distribution activities as they would be required to do if the activities in question were carried out by separate undertakings, with a view to avoiding discrimination, cross subsidisation and distortion of competition'. Initially, the Commission sought ownership unbundling instead of the mere unbundling of accounts. However, due to strong opposition, particularly from Germany and France, the 2003 directives neglected this issue. Both directives did address the rules for the access of third parties to transmission and distribution systems, though (EP/Council 2003b: Art. 20; 2003c: Art. 18). Thus, the member states were obliged to ensure that third parties could access transmission and distribution systems based on tariffs that would be applied to all eligible customers. Furthermore, the Community's competition rules had to be satisfied, ensuring that no party is discriminated against. Finally, the directives required the member states to establish independent regulatory authorities responsible for securing 'effective competition and the efficient functioning of the market' (EP/Council 2003b: Art. 23; 2003c: Art. 25). Despite these guidelines, however, the directives remained rather vague concerning the structure of the institutions to be installed. Consequently, the regulatory authorities in the member states differed with regard to their organization and regulatory practice. Moreover, personal relationships between energy companies, politicians, and member-state authorities continued to hinder the strict implementation of the Commission's targets.

The two directives were complemented by Regulation 1228/2003, which specified the 'conditions for access to the network for cross-border exchanges in electricity' (EP/Council 2003). In order to enhance such exchanges, the regulation included five principles, summarized by the Energy Community (2014) as follows:

1. Transmission system operators must be compensated for costs incurred as a result of hosting cross-border flows of electricity on their networks. This is a prerequisite for an open, competitive market.
2. Non-discriminatory and transparent tariffs for access to networks must be set to reflect payments and receipts resulting from compensation between transmission system operators. This is a precondition for effective competition in the internal market.

3. In cases of network congestion, the allocation of cross-border capacities shall be addressed with non-discriminatory, market-based solutions to give efficient signals to market participants and transmission system operators.
4. Different safety, operational and planning standards used by national transmission system operators should be harmonized in order to avoid distortion of competition.
5. Publication of relevant data for the market participants to eliminate asymmetries in information.

Despite member-states' obligation to implement Directives 2003/54/EC and 2003/55/EC by 1 July 2004, progress was slow and incomplete. In November 2005, the Commission criticized several member states, including Spain, Luxembourg, Greece, Estonia, Portugal, and Ireland, for only minimally implementing the directives and that their electricity and gas markets were still primarily organized by the state.

As a consequence, price differences between the member states persisted and cross-border trade of electricity and gas remained limited. Just a few, big national players continued to dominate the market, and they had no interest in informing their customers about how to switch to competing energy suppliers. Although major enterprises increasingly tended to choose their energy provider according to the best cost–benefit ratio, small companies and private households mostly stayed with the same company without comparing energy prices. Therefore, there was virtually no incentive for new businesses to enter the market. Despite two legislative packages, the member states achieved only uneven progress and almost no meaningful competition (European Commission 2007c).

The third legislative package

When competition cannot be guaranteed in one or several sectors, the Commission has the right to inquire about its causes. In terms of realizing the internal gas and electricity market, Council Regulation 1/2003/EC lays down specifically:

> In the course of that inquiry, the Commission may request the undertakings or associations of undertakings concerned to supply the information necessary for giving effect to Articles 81 and 82 of the Treaty and may carry out any inspections necessary

> for that purpose. The Commission may in particular request the undertakings or associations of undertakings concerned to communicate to it all agreements, decisions and concerted practices. (Council of the European Union 2003b, Art. 17)

The Commission has exercised this right, such as in 2005, and conducted a comprehensive investigation into the energy sector. In its report, the DG Competition concluded that the energy sector was plagued with areas where competition was not functioning well and needed to be addressed rapidly 'in order for liberalisation to bear fruit' (European Commission 2007j: 4). Specifically, it characterized many member states' markets as having high levels of concentration, cartel-like agreements, and inadequate unbundling of network and supply. It also identified insufficient or unavailable cross-border capacity, and thus insufficient market integration, and a lack of or delayed investment, as well as increases in gas and electricity wholesale prices that could not be explained in terms of higher costs. European Commissioner for Competition Policy Neelie Kroes summarized these points in January 2007, emphasising that '[...] more than a decade after having launched the drive for liberalization, we are still far from having a single, competitive and well-functioning European energy market' (Kroes 2007). She went on to announce further measures aimed at remedying the situation, among which were not only the rigorous application of competition law and the strict surveillance of the correct implementation of the second legislative package but also the recasting of existing regulations (European Commission 2007d). Consequently, on 19 September 2007, the Commission adopted a comprehensive third package of legislative proposals (European Commission 2007e–i). Similar to the previous ones, these also aimed at realizing competitiveness, sustainability, and supply security (the three pillars of comprehensive energy security) in the electricity and gas market.

In January 2009, the Council adopted a position on the Commission's proposal. It took only two months for the Parliament and the Council to agree on a common position. The EP formally adopted it in April, followed by the Council in June. It was then published in the Official Journal on 14 August 2009 (EP/Council 2009a–b, f–h) and came into force in September. The most important (and most controversial) aspects of that legislation can be captured by the words 'unbundling' and 'third-party access', and its immediate effect observed in the establishment of the Agency for the Cooperation of Energy Regulators (ACER).

The right of the EU to impel national companies to unbundle some aspect of their business is rooted in EU competition law and, on numerous occasions, the Commission successfully compelled companies to sell shares and, thus, unbundle their activities (Schumann 2003). The unbundling of vertically integrated companies was already discussed by the member states during the negotiations on the second energy package. Although some progress was reached during those negotiations, the Commission saw potential for optimization. The compromise negotiated between the Council and the EP in 2009 offered three options for the companies in the member states. The first was ownership unbundling as already proposed by the European Commission in 2007. The second suggested setting up an independent system operator (ISO), and the third included the introduction of an independent transmission operator (ITO). The ISO option implies that companies retain the ownership of their transmission networks, but that the independent system operator controls the day-to-day management. The third option, the ITO model, was based on a proposal made by Germany, France, and six other member states – governments that would have been able to block any decision in the Council. Their position was duly reflected in the final directive. As EurActiv (2009) reported, 'They obtained the right for former state monopolies – such as EDF and GDF in France and E.ON and RWE in Germany – to retain ownership of their gas and electricity grids, provided that they are subjected to outside supervision'. The selection of the ITO option exemplifies the significance of state actors in the EU energy policy process (see Chapter 4). Yet they were not the only actors to raise obstacles.

Both of the EU sector associations representing the electricity (EURELECTRIC) and gas (Eurogas) industries shared, in principle, the Commission's preference for ownership unbundling. However, EURELECTRIC raised its concerns about the introduction of further legal rules. They argued that the current legal framework was sufficient for creating a healthy internal energy market and that the Commission should, instead, ensure that Directive 2003/54/EC was implemented correctly in the member states. Specifically, they noted, 'The Directive has found the correct balance, allowing companies to retain ownership of their networks while putting in place strict rules to ensure the independence of network operators in relation to the networks they operate, maintain or develop' (EURELECTRIC 2007: 19). Eurogas shared the perspective that further unbundling would overburden the industry and therefore supported the argument raised by EURELECTRIC that the implementation of existing legislation should be the top priority (Eurogas 2006: 2).

The concerns raised by the sector associations reveal the trans-dimensional nature of what at first glance looks to be a primarily internal dimension problem, but in practice also involves the external dimension. At the time, the discussions on third-party access centred on the role of powerful non-EU energy companies in the EU market. Expert opinions on the issue ranged from completely prohibiting entrepreneurial activities of non-EU businesses in this area to a positive evaluation of the capital such companies invest in the EU market. According to the Third Energy Package, the national regulatory authorities are responsible for certification, which can be denied when a company does not fulfil the EU requirements concerning unbundling or threatens the member state's or the Union's energy supply security. Before a decision is made, however, the NRAs have to notify the Commission and seek its opinion on the specific case. One implication of these new rules is the dispute between the EU and Russia's Gazprom over unbundling in regard to the South Stream gas pipeline. As recently as April 2014, Gazprom was still tangling with the EU over its laws that ban suppliers from owning transit facilities (e.g. pipelines), claiming that such rules violate international rules (World Trade Organization 2014). In December 2014, the Russian President Putin and Gazprom CEO Miller announced a stop to the pipeline, citing EU regulations that placed unacceptable terms on the pipeline project.

Finally, the third issue dominating negotiations was Regulation (EC) No 713/2009 on establishing the Agency for the Cooperation of Energy Regulators (ACER) that was adopted to coordinate and complement the work of the national regulatory authorities. ACER was officially launched in March 2011. Its primary focus has been the development of EU-wide market rules. The agency was established on the basis of the already existing structures and functions of the European Regulators' Group for Electricity and Gas (ERGEG). The Commission repeatedly expressed its concerns about the ERGEG's progress, which had 'not resulted in the real push towards the development of common standards and approaches that is necessary to make cross-border trade and the development of first regional markets, and ultimately, a European energy market a reality' (European Commission 2007i: 9–10). The Commission especially criticized the consensual approach dominating the work within ERGEG, which required 'the agreement of 27 regulators and more than 30 transmission system operators to reach agreement'. This 'led to a number of non-binding codes and efforts to reach agreement on common approaches through "gradual convergence" but has not lead to real decisions on the difficult issues that now need to be taken' (ibid: 10). Thus, ACER was endowed with comprehensive competences.

Similarly, the national regulatory authorities were strengthened in order to emphasize the relevance of independent institutions in a sector traditionally dominated by state decisions. Together, these moves served to 'guarantee the independence of the regulatory authority' and 'ensure that it exercises its powers impartially and transparently' (EP/Council 2009b). The NRAs now cooperate with each other (and other authorities at the member-state level), providing insight into the relevant data. As the Commission now also plays an important role in steering the information exchange and can adopt guidelines determining the degree of cooperation, the third legislative package not only brought Europe closer to realizing its internal energy market but also strengthened the Commission's oversight of matters previously relegated solely to the member states and the external dimension.

Further elements of the third legislative package worth mentioning include the expansion of cooperation between transmission system operators and the establishment of both the European Network of Transmission System Operators for Gas (ENTSO-G) and the European Network of Transmission System Operators for Electricity (ENTSO-E). The latter had already been established in 2008 with the primary tasks to ensure that the electricity transmission network operates in a secure and reliable way, to promote investments, and to facilitate the interconnection of the European grid. The formalization of ENTSO in the third legislative package constitutes an upgrading of existing informal structures and, thus, a type of 'regulated self-regulation' (Berg 2001; Schulz and Held 2004).

Finally, in an indisputable act of consumer emancipation, the legislation strengthened both consumer protection and rights. European gas and electricity consumers are now guaranteed the right to receive a detailed overview of their consumption and possess the right to switch suppliers with only three weeks' notice. While such changes may seem trivial in relation to the grander tasks of unbundling the accounts of energy providers and distinguishing between distribution and transmission networks (targeted already in the first legislative package), they achieved a harmonization of consumer rights across the Union that provide a key component both for further integration of Europe's unified internal energy market, and even a prospective Energy Union, as well as the dissemination of EU energy law beyond its borders through the Energy Community (see Chapter 7).

The single energy market: Current state and challenges ahead

Despite a few notable steps towards an internal energy market, including the third legislative package, the EU still has a long way to go. Domestic market characteristics and prices still vary considerably between the member states, and substantial gaps in the Union's energy infrastructure limit the ability of the member states to rapidly respond to shortages or interruptions. Moreover, while most member states export some quantity of electricity annually (Figure 5.1), they also import large volumes of electricity (Figure 5.2) and natural gas (Figure 5.3).

In the electricity sector, for example, rather than operating a single Union-wide grid linking suppliers to consumers, the EU operates 28 national grids. These are divided into five regional groups (Continental Europe, Ireland, United Kingdom, Nordic, and Baltic) and two voluntary Regional Groups (Northern Europe and Isolated Systems), which are coordinated by The European Network of Transmission System Operators for Electricity (ENTSO-E). Despite the grid's complexity, the mechanism of energy demand and supply remain in place: power stations feed the amount of electricity demanded by consumers into the grid. Because national electricity demand swings daily, generators export or import surpluses at prices that have ranged from between €29 and €60 per megawatt-hour between 2010 and 2013, depending on the regional market (European Commission 2013p). In 2013, for example, slightly more than 1.2 thousand TWh were traded on the Leipzig-based European Energy Exchange's (EEX) Power Derivatives Market alone (EEX 2014), a figure that must be understood as only part of the overall trade in electricity, since much of it is traded over

Figure 5.1 *Electricity exports (TWh, 2012)*

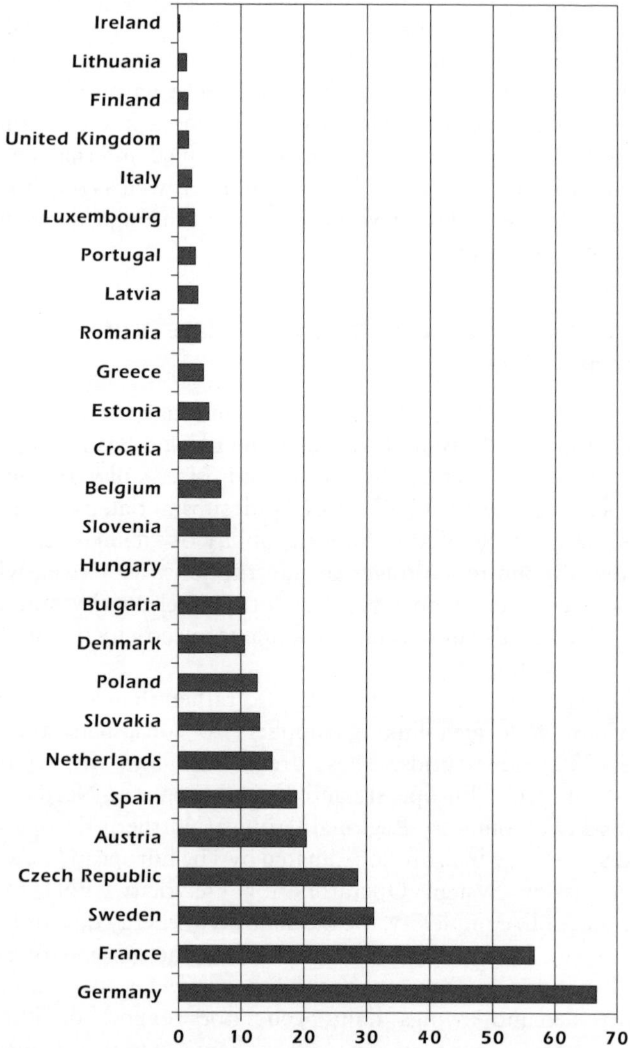

Source: Based on Eurostat data (Eurostat 2015b)

the counter and based on reciprocal trades. The amount of electricity traded varies between the member states, but seen in the context of the global electricity trade, the amounts are massive.

Figure 5.2 *Electricity imports (TWh, 2012)*

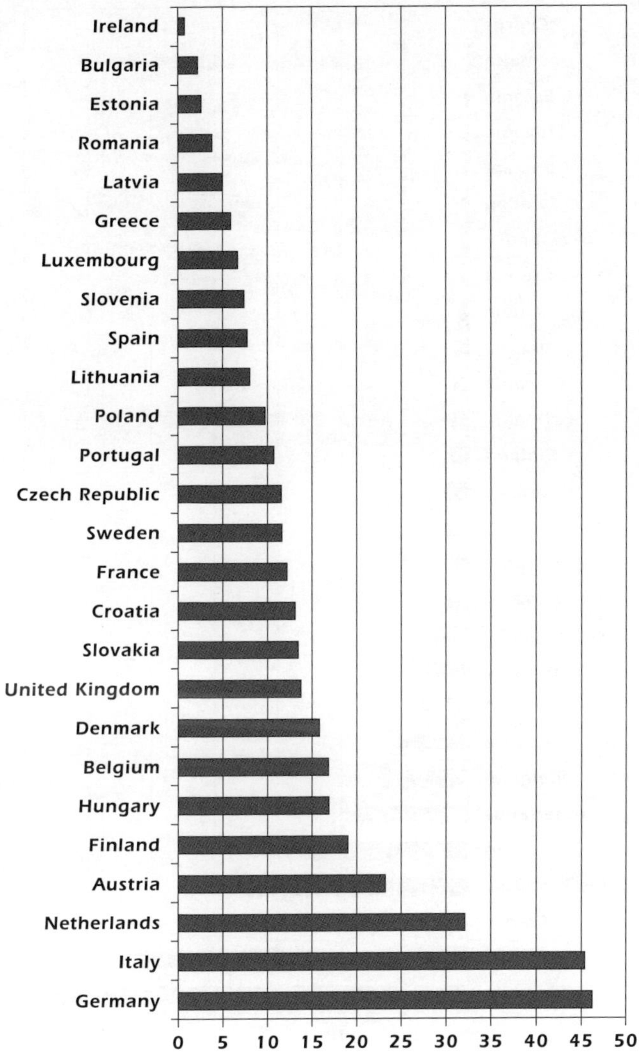

Source: Based on Eurostat data (Eurostat 2015c)

Germany exported 66.6 TWh in 2012, which was slightly more than double the amount exported by Sweden (31.28 TWh) and almost ten times as much as Belgium (6.911 TWh), and more than

Figure 5.3 *Gas imports (millions of terajoules)*

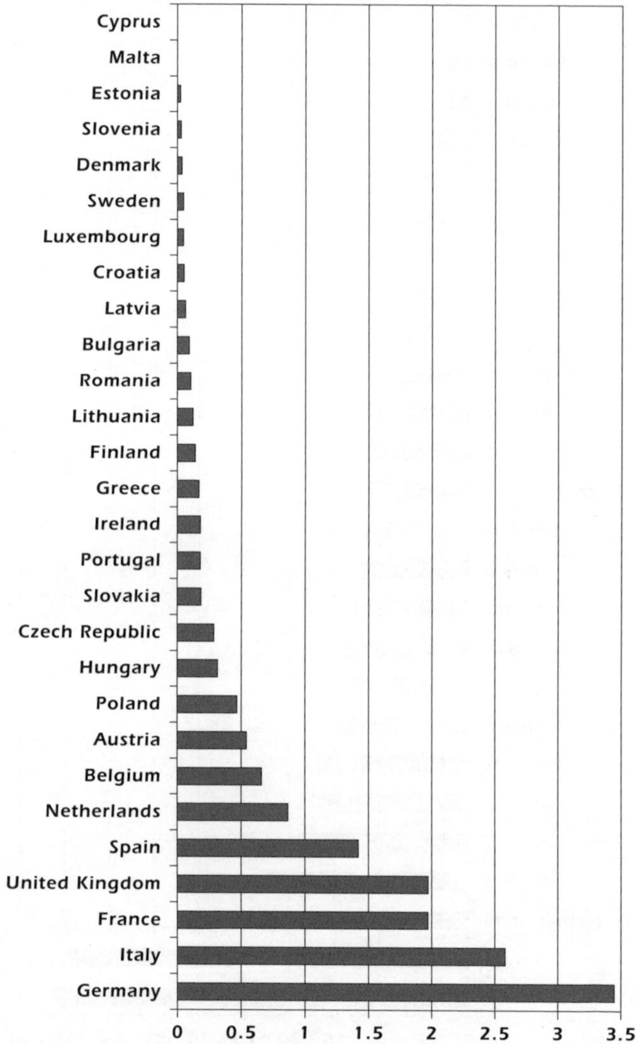

Source: Based on Eurostat data (Eurostat 2015d)

Canada (57.97 TWh), Russia (19.14 TWh), or the United States (12 TWh).

Looking at the electricity exports of Germany, one can also see the uneven penetration of renewables. As Germany generates a substantial amount of renewables that are by their nature uneven in their delivery (see Chapter 2, Section '*Storage, predictability, and negative externalities*'), German generators have to produce more electricity than they need nationally to stabilize the grid, creating surpluses that either have to be consumed or sold. This in turn depresses prices that hit the bottom line of generators. In the long run, the cycle of declining wholesale electricity prices runs the danger of either pricing renewables out of the market or forcing generators to cut back on production, neither of which will help the EU achieve its stated goal of comprehensive energy security.

Meanwhile in the gas market, indigenous EU production (159 bcm) accounted for the largest single source of gas for EU customers in 2012, making up almost a third of total supplies (Eurogas 2013: 6). Most of the production was consumed directly by the producing member states. However, the Netherlands exported substantial amounts to its fellow EU members; and others, such as Austria and Slovakia, re-exported large volumes of Russian gas. According to the International Energy Agency (2013a: 13), the Netherlands was the eighth largest net exporter of gas worldwide in 2012, while the EU's largest single supplier, Russia, led the world with 185 bcm in net exports, with Qatar (120 bcm) and Norway (109 bcm) not far behind. It is notable that three of the top five global exporters of natural gas, which totalled 342 bcm in 2012, accounted for 54% of the EU's net natural gas supplies of circa 470 bcm that same year (Eurogas 2013: 6).

Energy markets within the EU not only vary concerning their export and import structures; they also vary with regard to energy prices. For example, in 2013, average annual electricity prices (including taxes) for most households ranged from circa 0.09 €/kWh in Bulgaria to almost 0.30 €/kWh in Denmark (see Figure 2.1) while gas prices for households, which averaged circa 0.07 €/kWh, ranged from just under 0.3 €/kWh in Romania and slightly above 0.12 €/kWh in Sweden (see Figure 2.2).

In contrast to what the Commission expected from the establishment of the internal market, electricity prices paid by companies and households have actually increased over the last 20 years. As recently as January 2014, the Commission was still arguing that '[a] fully integrated and competitive energy market could result in cost savings of between €40-70 billion up to 2030' (European Commission 2014: 9). Accounting for the disparity between its expectations and reality, the Commission blamed, among others, high national taxes, fuel costs, limited competition, and state intervention for the price rise. Rising energy costs were among the items discussed by the 160 stakeholders participating at the Berlin Energy Forum in February 2014 (European Commission 2014n). Given that the purpose of the forum was to gain input from stakeholders in order to debate the energy sector objectives up to 2050, it was not surprising that its participants found that the best response was to coherently apply internal market rules, provide stable investment conditions, and to diversify energy sources. While such recommendations fit EU stakeholder strategy, it is unlikely we will see a levelling of energy prices across the member states any time soon. Indeed, energy prices are increasing uniformly across all member states (European Commission 2014f). While this may not fit neatly with the Commission's plans, it at least indicates that the European energy markets are moving in unison.

As outlined at the beginning of this chapter, the Commission expected that the internal energy market would not only result in lower consumer prices, but also attract higher investments in infrastructure projects. The maintenance and expansion of grid and generation infrastructure is essential for the proper functioning of national economies and the provision of appropriate living conditions for the member states' citizens. The international energy crises of the last decades, especially the gas dispute between Russia and Ukraine, renewed public awareness of the urgent need to invest in cross-border infrastructure to secure energy supplies, particularly in times of supply disruptions. Modernizing cross-border infrastructure thus quickly became a major issue on the political agenda.

In April 2013, the EU identified several high-priority trans-European infrastructure projects in the areas of gas and electricity, which it expected would ease transportation, mitigate exogenous supply shocks, and increase competition between energy producers and suppliers (EP/Council 2013a). Among these were the North Seas Offshore Electricity Grid (NSOG), the Southern Gas Corridor (SGC), the Baltic Energy Market Interconnection Plan in gas ('BEMIP Gas'), two additional North–South interconnections (respectively, in Western Europe (NSI West Electricity and NSI West Gas) and Central-Eastern and South-Eastern Europe (NSI East Electricity and NSI East Gas)), and the Baltic Energy Market Interconnection Plan in electricity ('BEMIP Electricity'). Supporters of expanding the EU's energy infrastructure note that adding cross-border interconnections will enhance reliability and thus security of supply, increase competition, facilitate increased exchange, and maximize the use of profitable locations to generate electricity and import gas. One backer, Spanish Secretary of State for Energy Alberto Nadal Belda, argued that such 'interconnections are crucial for the Internal Energy Market and a true added value for the interests of the EU's industry sector' (European Commission 2014d: 12–13).

Political support notwithstanding, building these interconnections will be expensive, and despite its steadfast support for market liberalization, the Commission apparently doubts that the market will be able to deliver the full amount of the required investments due to 'difficult access to finance and lack of adequate risk mitigating instruments' (European Commission 2011i: 11). Based on a 2011 estimate that the upgrades required to meet its projected needs and goals could cost upwards of €1 trillion by 2020, a fifth of which alone would be required to modernize existing transmission networks, the Commission confirmed in 2012 that 'although investment is being made in all sectors, it is not reaching the rate needed to meet the policy ambitions' (European Commission 2012d: 4). In October 2013, the Commission adopted a list of 248 projects of common interest (PCI), which had been selected by twelve regional groups according to the guidelines for trans-European energy infrastructure (TEN-E). In order to be included in this list, a project has to contribute to market integration, competition, energy supply security, or the reduction of industrial emissions in at least two member states. Although the Commission does offer financial support to projects that have been labelled as 'projects of common interests' (PCI) through its €5.85 billion Connecting Europe Facility (CEF), obstacles to private investment continue to hamper progress. An analysis of existing and planned EU gas and electricity infrastructure capacity indicates that renewed impulse at the European level will be required to secure increased investments in energy infrastructure in the future (Table 5.1). It seems clear that more will have to be done to overcome investment barriers in the member states. In what now amounts to a prudent act of foresight, the Commission had already begun to address that point in 2010 by proposing to establish a framework for the exchange of data and information on energy infrastructure projects based on Council Regulation 617/2010 (EP/Council 2010b), and passed a new regulation replacing the former in February 2014 (EP/Council 2014).

The limited progress made on the infrastructure front is indicative of the difficulties faced by the Commission in matching its goals of concurrently achieving a functioning internal energy market

Table 5.1 *Existing and planned capacity in electricity and gas infrastructure (excluding LNG)*

	Gas pipelines (km)	Cross-border gas pipelines (To) (mcm/ day)	Cross-border gas pipelines (From)	Gas storage (bcm)	Electricity generation (MW)	Electricity transmission (km)	Cross-border electricity transmission (MVA+MW)
Existing (as of 1 January 2011)	125,027.6	2,467.8	2,315.2	95.6	839,463.3	261,097.2	237,836.4
Planned additions* (as of 31 March 2011)	5,734.7	390.9	356.7	25.5	87,229.5	28,113.9	72,945.7

Note: *Planned additions correspond to the sum of underconstruction and planned projects, minus infrastructure that will be decommissioned

Sources: European Commission (Notifications under Council Regulation 617/2010 and Commission Regulation 833/2010) ENTSO-G, ENTSO-E, Platts

based on liberal principles with the realities of national interests and private sector investment strategies. Part of the problem remains state involvement in energy companies, traditionally considered an obstacle to the proper functioning of the internal energy market. However, even the Commission admits that a certain degree of state intervention may be necessary to achieve its goals. Although principally cautious about state intervention, the Commission now recognizes that in some sectors, such as renewables, state aid may be provided at the national, regional, or local level so long as it is well-designed and properly coordinated at the EU level (European Commission 2013j).

Faced with increasing efforts by some member states to offer far-reaching public support for investments in modern generation capacity, the Commission announced its intention to adopt new guidelines on environmental and energy aid for 2014–2020 (2013k). Following a two-month public consultation that began in mid-December 2013, the Commission issued a Communication in June 2014 that, alongside accentuating its commitment to support cross-border energy infrastructure projects, represented significant compromises. For example, arguing that its goal was to promote 'a gradual move to market-based support for renewable energy', it recommended introducing a competitive bidding process for public support and feed-in premiums to replace feed-in tariffs in order to gradually expose renewables to market signals. It also accepted the right of states to provide public funding in cases where there is a risk of insufficient electricity generation capacity and proposed to simplify procedures to support projects and companies in the fields of energy and the environment (European Commission 2014i). While not abandoning its long held principle of liberalizing Europe's diverse energy markets, such steps seem to represent a certain degree of prudent backtracking, a reality check of sorts for the concomitance of the three pillars of comprehensive energy security.

Concluding remarks

The EU's internal energy market policies have only been partially successful. Although considerable institutional and procedural reforms were introduced, their scope and implementation at the national level remains insufficient. The failure to enforce ownership unbundling to guarantee grid independence instead of a mere legal unbundling is only one example. Member states are still lagging behind on the implementation of EU legislation. In a November

2012 Communication, the Commission criticized the member states as being 'slow in adjusting their national legislation' and often pursuing 'inward-looking or nationally inspired policies', both of which hampered the effectiveness of the adopted policy measures (European Commission 2012d). Nevertheless, the European Council decided to set 2014 as the deadline for the full completion of the internal gas and electricity markets (European Council 2011). In February 2014, the Commission announced an important milestone in that pursuit when electricity grid operators and power exchanges from 14 EU member states (Belgium, Denmark, Estonia, Finland, France, Germany, Austria, UK, Latvia, Lithuania, Luxembourg, the Netherlands, Poland, and Sweden) joined Norway in inaugurating a pilot project for one-day-ahead market coupling (European Commission 2014e) and, as of May 2014, the Commission was working on a regulation to make the practice of market coupling binding for all member states. Such efforts notwithstanding, the EU in mid-2014 still lacked the fully integrated electricity and gas markets that it deems vital to a functioning internal energy market; and it remains to be seen whether those markets will deliver the expected results when they ultimately come to fruition.

Chapter 6

Climate Change, Energy Efficiency, and the Quest to Expand the Use of Renewable Energy Sources

In recent years, the Commission increasingly emphasized the importance of finding 'cost-efficient ways to make the European economy more climate-friendly and less energy-consuming' (European Commission 2015a). To that end, responsibility for all climate-related topics previously held by the DG for Environment was assigned in February 2010 to a DG for Climate Action. In 2014, Connie Hedegaard, then Commissioner for Climate Action, noted that the ambition of the EU member states in realizing these targets should serve as a motivation for other countries to similarly aim for environment-friendly economic growth (Hedegaard 2014). Hedegaard's comments are indicative of how climate actions stand at the crossroads of internal and external policymaking. It is internal insofar as it relates to the regulation of energy efficiency standards and the promotion of renewable energy forms in the EU. The external dimension applies to the EU's international obligations and its claim to fame as a united political force on the world stage. The EU's external climate policy agenda was clearly driven by the entry into force of the United Nations Framework Convention on Climate Change (UNFCCC) in March 1994 (Oberthür and Pallemaerts 2010). The Convention 'sets an overall framework for intergovernmental efforts to tackle the challenge posed by climate change' (UNFCCC 1994) and has since been ratified by 195 countries (UNFCCC 2015). We identified climate change policies as multidimensional (see Chapter 1) precisely because the goals set within its context are unachievable without specific internal regulations, particularly increases in efficiency, reductions in the use of carbon-intensive fuels, and international reciprocity.

Although there existed a general consensus in the EU on the need for action to mitigate global warming during the early days of the EU's climate policy, the policy process was dominated by differences between the EU member states on the appropriate instruments. These differences resulted in the member states' inability to adopt binding targets for the participating countries: 'The lack of clear mitigation commitments in the UNFCCC resulted in the failure to establish effective GHG mitigation policies at the EU level [...]' (Oberthür and Pallemaerts 2010: 53). Today, the EU understands itself very much as an international agenda-setter in the area. According to the DG for Climate Action (European Commission 2014o), the EU has been a 'driving force in international negotiations on climate change' and has made a major contribution towards a new global climate agreement. It has since changed that description, now placing itself 'at the forefront of international negotiations for a new global climate agreement' (European Commission 2015b). Despite this slight change, opposing voices, especially among scientists, emphasize that the EU is losing its impact in global climate change negotiations. Indeed, there has been a notable lack of progress since the United Nations Climate Change Conference met in Bali in 2007 and adopted measures aimed at reaching a binding agreement to succeed the Kyoto Protocol by its Copenhagen Summit meeting in late 2009 (known as the Bali Roadmap; UNDP 2008, UNFCCC 2007). Progress has since been slow and convoluted. The Copenhagen Summit failed to deliver on its promise. The resulting Copenhagen Accord produced neither a legal treaty nor a target year for peaking emissions, illustrating 'a worrying trend' (Fernandez Martin 2012: 193) that the EU was losing its status as an environmental champion.

This chapter provides a state-of-the-art snapshot of current EU climate policies and future goals by examining the EU's measures to increase the share of its renewable energy sources, to strengthen energy efficiency, and to introduce new technologies and materials such as carbon capture and storage (CCS), nuclear fusion, large-scale wind, and concentrated solar power (CSP) plants.

Increasing the share of renewable energy sources

One of the first initiatives concerning the promotion of the share of renewable energy sources was the Commission's White Paper *Energy for the Future: Renewable Sources of Energy* issued in

November 1997 (European Commission 1997). This initiative was based on three arguments. First, replacing fossil energy sources should contribute to reducing carbon dioxide emissions resulting from the use of oil, natural gas, and coal. Second, increasing the share of renewable energy was expected to reduce the Community's dependence on energy imports from other countries by utilizing domestically available sources of energy for electricity generation. Third, by promoting renewable energy sources, the demand for innovation should be increased, strengthening the regional economy and providing new employment opportunities (Howes 2010) (i.e. introduce a new industrial age commonly referred to either as a green or low-carbon economy). According to the Commission, renewable energy sources were 'unevenly and insufficiently exploited in the European Union' (European Commission 1997). Therefore, the White Paper set the goal of doubling the share of energy provided by renewable sources to 12% by 2010. In so doing, the Commission laid the groundwork for a cause–effect policy model (increased use renewables equates to reduced GHG output) that would serve as the basis for its future climate change policies and lead to multiple rounds of increasing renewable targets (European Commission 2008a, b, 2011a–c, 2013f).

In March 2006, the European Council pointed to the need of assigning the EU a leading role in combating climate change and asked the Commission to develop an action plan on how to promote renewable energies in the long term. In so doing, climate change was used as a political tool to centralize energy-related issues traditionally restricted to the sole authority of the individual member states. This is also reflected in the *Renewable Energy Road Map* (European Commission 2007m), which emphasized the value of renewable energy sources for a sustainable future: 'They [renewable energy sources] are largely indigenous, they do not rely on uncertain projections on the future availability of fuels, and their predominantly decentralised nature makes our societies less vulnerable' (European Commission 2007m).

Based on these assumptions, the Commission proposed in January 2008 its 20-20-20 targets (European Commission 2008a, b), which included: (a) increasing the share of renewable energy in total EU energy consumption to 20%, and the share of biofuels in transport to a binding minimum target of 10%; (b) cutting greenhouse gas emissions by at least 20% (below 1990 levels); and (c) reducing energy consumption by 20%. In order to achieve the overall 20% target in the renewables sector, Directive 2009/28/EC of 23 April 'on the promotion of the use of energy from renewable sources' (EP/Council 2009d) required the member states to adopt national action plans indicating the measures intended to realize the Commission's goals. However, in a clear indication of the perennial problem of harmonizing member-state interests, particularly in relation to their national energy, the individual targets each member state had to fulfil varied substantially, ranging from 10% in Malta to 49% in Sweden (EP/Council 2009d: Annex 1). Figure 6.1 lists the national overall targets for the share of energy from renewable sources in gross final consumption of energy for 2020, and compares these to the share of renewables in consumption in 2004 and 2013. As the data shows, substantial changes are still required in France, Germany, and the United Kingdom, three of Europe's most important industrial economies.

In order to increase the share of renewable energy in energy consumption, the Commission consistently draws special attention to the transport sector. The Renewable Energy Road Map (European Commission 2007m) called for not only a general increase in renewables but also an overall reduction in annual fossil fuel consumption, by over 250 Mtoe, by 2020. Touching upon all three dimensions of the energy policy typology, it described the advantages of replacing fossil fuels in terms of, for example, a reduced dependence on imports from non-EU countries (external), new investment opportunities in the renewable energy sector (internal), decreased CO_2 emissions, and improved air quality (multidimensional). However, especially with regard to the Commission's goal to increase the amount of biofuels in transport, conflicting interests between a wide range of actors, including scientists and politicians, prevented sustained success. The central scientific critique (e.g. IEA 2008, OFID 2009) emphasized that the production of biofuels negatively affects other industry sectors, such as agriculture, private consumption, and health. Another argument raised against biofuels is that they were not necessarily more environment-friendly than fossil fuels, because of the huge amount of energy necessary to produce them, which could, and probably does, potentially increase CO_2 emissions rather than reduce them (Howarth et al. 2009, Pimentel and Pimentel 2007: 269).

Figure 6.1 *National targets for the share of energy from renewable sources in gross final consumption of energy in 2004, 2013, and 2020*

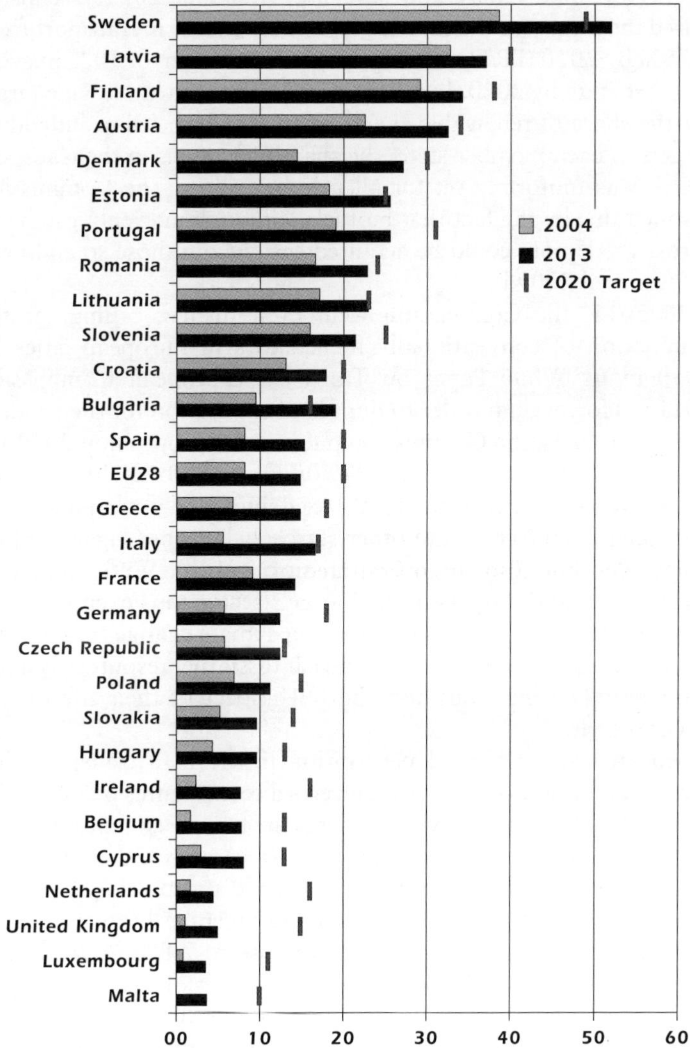

Sources: EP/Council (2009d: Annex I), Eurostat (2015g–h)

In response to these critiques, the EU passed Directive 2009/28/EC (EP/Council 2009d) 'to ensure [only] the use of sustainable biofuels' that 'generate a clear and net GHG saving without negative impact on biodiversity and land use' (Eurostat 2015e). Thus, it raised the target for the share of renewable energy in transport from 5.75% by 2010 (EP/Council 2003a) to a minimum of 10% in every member state by 2020. In contrast to the overall Union-wide target for the share of renewable energy sources, which led to individual targets in each member state, the share of biofuels in the transport sector was uniformly set for all member states. The Commission justified this by the fact that biofuels, although unevenly produced across the Union, could be acquired easily from those so endowed (EP/Council 2009d).

In 2011, the Commission went even further, calling for the elimination of conventionally fuelled cars in European cities by 2050 in its White Paper on Transport (European Commission 2011c). However, in order to limit land conversion for the production of biofuels, the Commission published a proposal in 2012 for a directive amending Directive 98/70/EC and 2009/28/EC, limiting the share of energy from food-based biofuels – meaning those produced from 'cereal and other starch rich crops, sugars and oil crops' – to 5% (European Commission 2012b). While unique in their use of land, targets in the biofuels sector thus exemplify one of the main complications in achieving climate change targets: that the practical fungibility of carbon-rich fossil fuel resources is questionable when one considers the full impact of their use on the environment.

An analysis of biofuel penetration in the EU's transportation sector reveals a mixed bag of uneven success (Figures 6.2 and 6.3). While the share of renewable energy in fuel consumption in the transport sector has risen (Eurostat 2014) since 2004, Europe's road to increased use of biofuels has been disproportionate among its members (Figure 6.2). In 2004, the average biofuels market share in the EU was only 1.0%, but rose to 2.8% in 2007, and then to 5.1% in 2012. While the percentages of market penetration are not high in absolute terms, the growth rates have been substantial where biofuels have found a reliable position in the market. Between 2004 and 2012, the share in Sweden rose from 3.8% to 12.6%, and in the UK it grew from 0.2% to 3.7%. This growth has not been the case everywhere, and there has even been a decline in some countries in recent years. In Cyprus, for example,

Figure 6.2 *EU28 renewable energy share in fuel consumption by transport (2004 and 2012)*

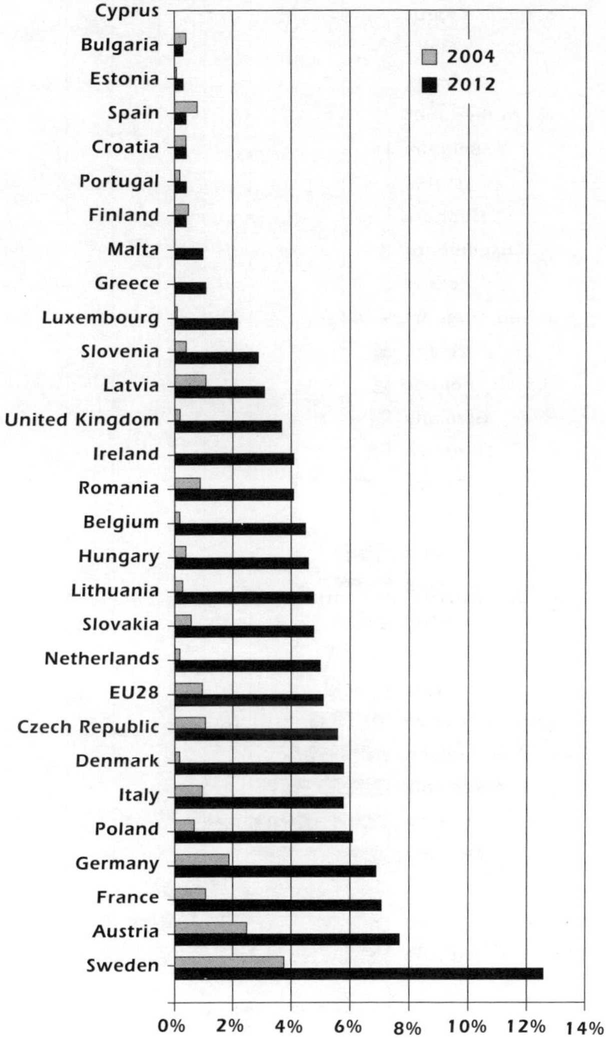

Source: Based on Eurostat (2014a) data

there were virtually no biofuels on the market in 2004. Their share peaked at 2.0% just five years later in 2009, but then dropped again to nil in 2011.

Figure 6.3 *Primary hydroelectricity production share of electrical energy available for final consumption in the EU (2012)*

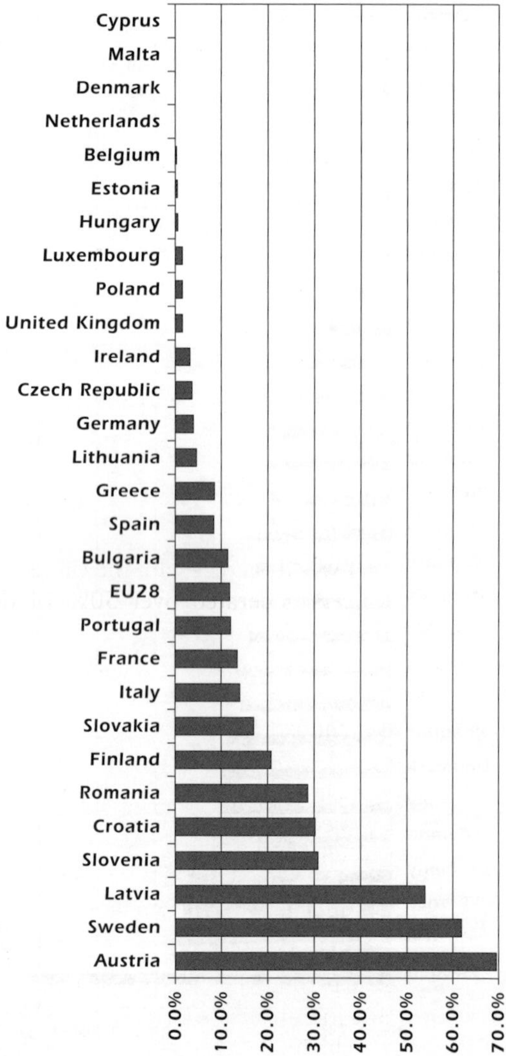

Source: Based on Eurostat (2014b, c) data

In addition to biofuels, the Commission sees significant potential in hydropower (and particularly small hydropower systems) to optimize electricity production and react to fluctuations in energy demand. Hydropower was the source of approximately 16% of global electricity generation in 2011, and the EU27 consumed approximately 9% of the world's hydroelectricity (European Commission 2014j: 16, IEA 2012: 7). In 2012, EU-based hydropower facilities constituted 15.6% of all EU-installed electricity capacity (almost 149 GW) and almost 46% of all of its renewable sources of electricity at 366 TWh, an output level that has held fairly steady since 1990 (European Commission 2014j: 88–90). Again, analysis of the data reveals a huge difference between the member states, with some producing no hydropower at all, while others such as Austria that in 2012 generated almost 70% of its electricity from hydroelectric sources (Figure 6.3).

Hydropower is a vital and reliable source for countries such as Austria, Latvia, and Sweden. However, it is of little use to a country in short supply of flowing water, such as Malta or Cyprus. Moreover, Europe's total hydroelectric capacity is relatively small when compared globally. The world's largest producer of hydroelectric power is China, which generated 699 TWh in 2011, or 19.6% of the world's total production that year, followed by Brazil's 428 TWh and Canada's 376 TWh. Together with the United States' 345 TWh, those four producers generated over 50% of the world's hydroelectric power (IEA 2013a: 19).

Beside its focus on biofuels and hydropower, the Commission considers wind power to be a highly promising renewable energy source. The wind power industry is a rapidly growing industry in the EU, making up 'around one third of all installed electricity generating capacity in the EU' since 2000 (European Commission 2014p). According to projections by the Commission, in cooperation with the European Wind Energy Association (EWEA), given sufficient support from all levels, 'onshore wind will be the largest contributor to meeting the 34% share of renewable electricity needed by 2020 in the EU, as envisaged by the 2009 Directive' (EWEA 2009: 6). Unsurprisingly, in order to realize this goal, EWEA logically calls for long-term EU investment in related technology and policy research, such as the North Seas Offshore Electricity Grid (NSOG) (see Chapter 5).

Similar to other renewable technologies, wind power technology is not free from questions. There are serious disadvantages and negative externalities as well. The most important ones include the unreliability of wind speeds, the high costs of related infrastructure, and the resulting environmental impacts, for example, from the removal of vegetation in places where plants are being built. Also relevant are the negative effects on the living conditions of populations located in the vicinity of wind farms, particularly noise and visual pollution, and reduced property values. The Commission recently emphasized that in order to continually increase the share of electricity generated from wind power in EU consumption, wind turbines have to move offshore. In early 2014 wind farms under construction were poised to add another 4,900 MW to the EU's overall installed wind power capacity. In the first six months of 2014 alone, more than 220 new offshore wind turbines were installed, with a combined capacity of 781 MW, and another 310 were awaiting connection to the grid (EWEA 2014). Despite these latest efforts, however, offshore wind farms have as yet contributed very little to meeting the EU's energy demands (Table 6.1), and given both the enormous investments required to build up wind energy resources and the long-term impact of forcing generators to overproduce electricity to ensure grid stability, there remains serious scepticism about their long-term economic viability. As Table 6.1 shows, wind energy generated only 205.8 TWh of gross electrical energy in the EU, or roughly a quarter of all renewable-based electricity, in 2012. Meanwhile, renewables accounted overall for only a quarter of the EU gross electricity generation, while nuclear power accounted for little more than another quarter. Just under half came from conventional thermal sources, including 18.7% from natural gas and another 27.4% from hard coal and lignite.

This continued concentration of carbon-rich fuels in the EU's energy mix reveals the limited progress made by the member states on the climate change front when it comes to power production. While the use of coal across the Union has certainly declined since 1990 (from 39%), as did the share of nuclear (down from 30.6%), and renewables have largely taken their place (up from just 12.6%), several member states, including some of the most adamantly pro-environment such as Germany, continue to rely heavily on coal. Indeed, the mix of energy resources used for electricity generation varies significantly across the Union (Figure 6.4). For example, in 2013, both Austria (60.5%) and Latvia (62.5%) generated more than half of their electricity production using hydropower. However, half of Slovakia's (55.3%) and Belgium's (51.3%) electricity came from nuclear power stations, which also accounted for 75.5% of France's electricity generation. Meanwhile, Malta (100%) and

Table 6.1 *EU28 gross electricity production, 2012 (in % and TWh)*

	Share	*TWh*
Renewables	24.2%	798.7
- Wind	6.2%	205.8
- Hydropower	11.1%	366.4
- Geothermal	0.2%	5.8
- Tidal/Wave	>0.1%	0.5
- Biomass	4.5%	149.4
- Solar	2.2%	71.0
Nuclear	26.8%	882.4
Conventional thermal	49.0%	1,614.1
- Solid Fuels	27.4%	901.8
- Gases	18.7%	614.7
- Petroleum	2.2%	72.5
- Non-RES Wastes + others	0.7%	25.1
Total	100.0%	3,295.2

The total figure for renewables of 798.7 is that quoted in the EU report. However, the actual sum of the individual totals is actually 798.9.

Source: European Commission (2014j: 91)

Cyprus 95.8%) rely almost exclusively on oil-fired power stations for their electricity.

Part of the problem in achieving the ambitious targets adopted in Brussels stems from the fact that the Commission can neither mandate thc actual energy mix of resources used in the production of electricity in its member states nor can it manage the exploration and development of its own energy resources. Unlike the world's other major energy producers and consumers, the EU has no centrally managed public lands upon which it can coordinate offshore drilling or mining – a fact that greatly complicates the

Figure 6.4 *Breakdown of electricity generation by source, 2013 (in %)*

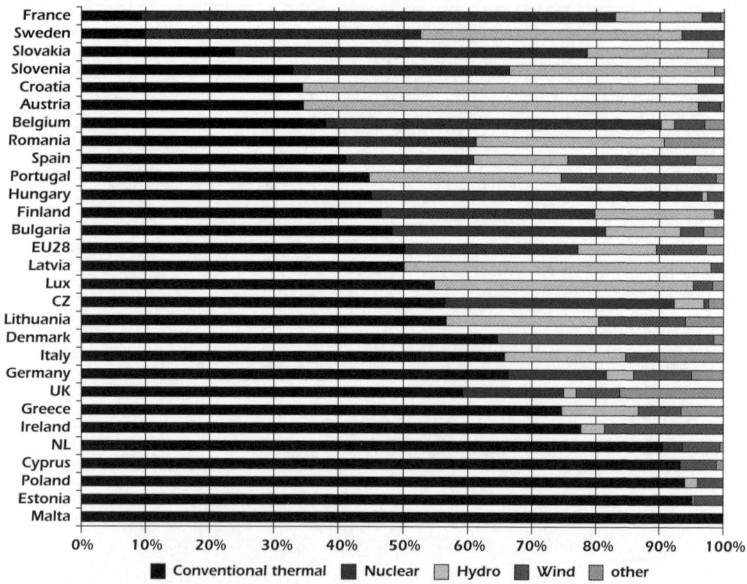

Source: Based on Eurostat (2014b, d, e) data

coordination of policies, particularly with respect to increasing or reducing domestic fossil fuel production. Yet even if the EU could centrally control national energy mixes, mining, and exploration activities, there is no guarantee that it could overcome all the other obstacles associated with those activities, or prevail over the complex set of political and economic interests of the many actors (see Chapter 4) involved the EU's energy sector. Hence, despite the progress in some areas, the overall trends in the renewable sector have to be assessed cautiously.

Based on a conservative evaluation, the share of renewable energy sources is projected to decline in comparison to other energy sources because economic crises, administrative and infrastructural barriers, policy shifts, and support-scheme disruptions repeatedly hamper investment in the sector (ECOFYS et al. 2012). For example, production capacities in the biofuels market have been insufficiently exploited in recent years, reducing the impetus for further investment. As one related report notes, in 2012, biofuel production in the EU was operating at 40% of capacity (bioethanol hovered between 50% and 60%), adding that the 'unused capacity' indicates 'that there is sufficient conversion capacity available for several years to come' (ECOFYS et al. 2012: iv). One reason for the failure to substantively increase the share of biofuels in the European transport market is that while the EU leadership sets targets for biofuels and legislates fuel-blend percentages in the consumer market, it simply cannot control how much the member states will invest in the requisite infrastructures. Despite all its achievements, the EU remains constrained by its own design.

So how does the EU plan to meet its ambitious goals? How will it ensure policy coherence across the member states? The Commission certainly is aware of the challenges responsible for the insufficient growth of the renewables sector and plans to compensate for them through increased investment in research and development and the distribution of revenues from the trade of emissions allowances (European Commission 2013d: 7). Some obstacles, however, will remain insurmountable without the full cooperation and commitment of the member states, which retain sole responsibility over key decisions about investments, licensing, and resource mixes. According to the Commission, progress in removing these barriers has been mediocre at best, and some member states have not even addressed the way in which they intend to make the necessary reforms.

In the event that EU law continues not be implemented correctly, the Commission can be expected to launch infringement proceedings against the respective member states, which it did in 2013 towards Belgium and Estonia (MEMO/13/470) and Italy and Spain (MEMO/13/820), each of which was requested to ensure full compliance with Directive 2009/28/EC. Likewise, the Commission referred Poland and Cyprus to the European Court of Justice for not correctly enacting the Renewable Energy Directive (IP/13/259).

The onus, however, cannot entirely be placed upon the member states. It would be remiss not to critically discuss the efforts initiated at the political level to increase the share of renewable energy sources in EU's overall energy mix. Renewables are expected to reduce import dependence and mitigate the risk of future oil price shocks. However, they simply are not predictable as an electricity source, at least in comparison to any fossil fuel. As one distinguished energy economist describes it, the costs resulting from this insecurity are immense (Helm 2012).

The EU's efforts to strengthen energy efficiency

Increasing energy efficiency as a means to reduce growth in consumption, particularly, but not exclusively, in terms of electricity, is a fundamental component of EU energy policy. At the macro scale, energy efficiency is measured as energy intensity, which measures the energy consumption of an economy by dividing the gross inland energy consumption (coal, electricity, oil, natural gas, and renewable energy sources) by gross domestic product (GDP). Energy intensity (EI) relates inversely to energy efficiency: the lower the intensity, the more efficient is an economy's consumption of energy.

Total energy intensity decreased in all EU28 countries between 1995 and 2012 (Figure 6.5). Yet, while the EU as a whole reduced the energy intensity of its economy by an average of 1.65% per year, individual member states' performances varied significantly. Energy intensity declined annually for most member states (Figure 6.6), on average between 1% and 4% (e.g. UK (–2.6%), Germany (–1.7%), and France (–1.1%)), but hardly declined at all in a few (e.g. Austria (–0.7%) and Italy (0.58%)), and two, Romania and Lithuania, saw substantial reductions. Averaging figures in this manner also conceals occasional annual increases in energy intensity. Such was the case for many member states that saw their EI increase in 2010 from the previous year, a phenomenon linked to the economic crisis of the preceding years and the subsequent recovery (Bosseboeuf et al. 2013).

The Commission regards the increase of energy efficiency as a central mechanism to realize its overall energy policy goal of comprehensive energy security (sustainability, energy supply security, and competitiveness). This has been quite explicitly reflected in the Commission's proposals of the last decade. Already in 2000, the Commission had published an action plan to reduce the amount of energy consumed in the Community (European Commission 2000b) and in June 2005, the Commission published the *Green Paper on Energy Efficiency* (European Commission 2005). According to that plan, the EU could reduce energy consumption by 20%, reduce greenhouse gas emissions and dependence on energy imports, and create jobs in the renewable energy sector. The Commission then asked stakeholders interested in participating in the policy process to submit their positions in the course of an open consultation. In so doing, the Commission hoped to identify how it

Figure 6.5 *Energy intensity (EI) of the economy 2000 vs. 2012. (Base 1995 = 100; gross inland consumption of energy divided by GDP (kg of oil equivalent per 1,000 EUR))*

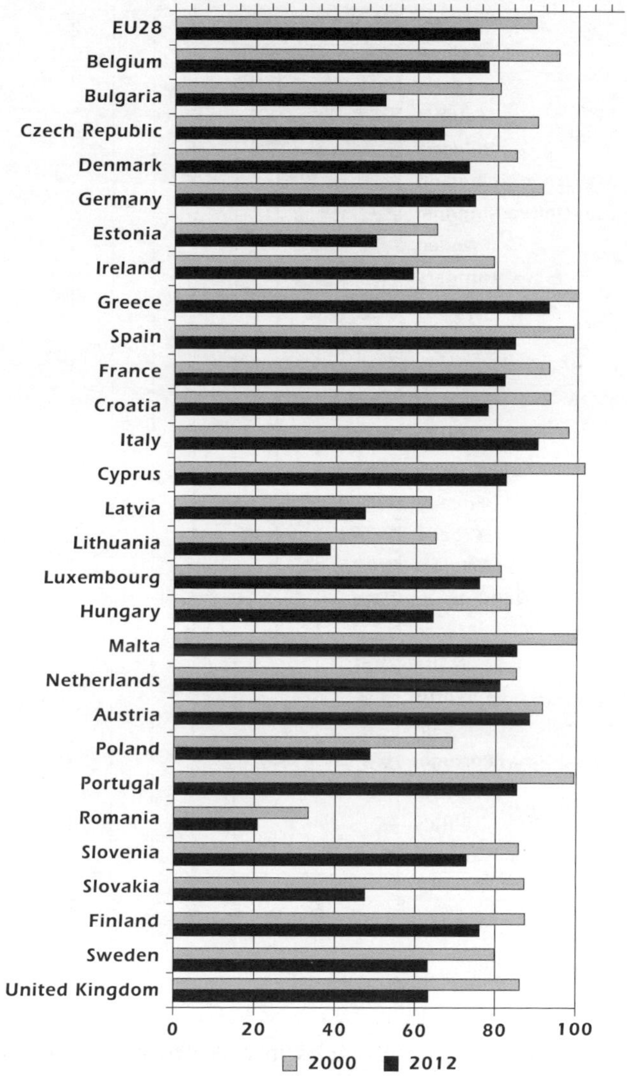

Source: Based on Eurostat (2014f, g) data

Figure 6.6 *Energy intensity, average annual change (1995–2012)*

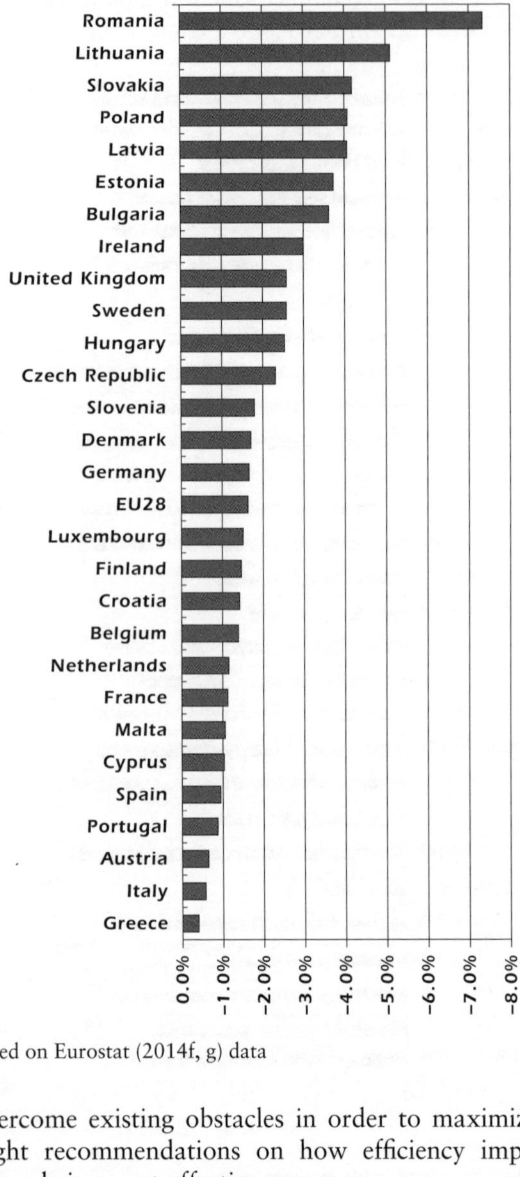

Source: Based on Eurostat (2014f, g) data

could overcome existing obstacles in order to maximize savings, and sought recommendations on how efficiency improvements could be made in a cost-effective manner.

The Commission envisaged huge potential in the buildings sector, where it expected to save energy by introducing energy performance certificates for all buildings exceeding 50 m^2 at their time of construction, sale, or rent, and even considered applying those measures to buildings under major renovation (European Commission 2005: 21ff). A second measure that concerned lighting worked on the premise that energy-saving light bulbs would use significantly less energy than standard ones (ibid: 22), a view that was supported in two meetings held by the Ecodesign Regulatory Committee in 2008. The latter resulted in calls for a draft regulation on how to improve energy efficiency in households and led the Commission to adopt two additional regulations requiring the eventual, total phase-out of the incandescent light bulb in Europe and replacing it with energy-saving alternatives by the end of 2012 (European Commission 2009b, c; see Chapter 4). The Commission also proposed certain requirements for heating, cooling, and electric motors in households. Finally, referring to the growth in the number of private cars and motorcycles on European roads (a by-product of Europe's economic success), it identified one of its most important goals moving ahead was to achieve substantial savings in the fuel consumption of vehicles, hence reducing overall energy consumption and CO_2 emissions in the transport sector (European Commission 2005: 23).

The action plan of 2006 (European Commission 2006b) identified further areas in which to save energy. In addition to reducing consumption in the construction and the transport sectors, the Commission proposed:

- to generally develop guidelines on improving energy end-use efficiency in all sectors;
- to better inform consumers on efficiency standards in order to transform the market;
- to offer economic incentives for energy efficiency investments, especially for small and medium enterprises;
- to review the modes of energy taxation currently applied;
- and to establish international partnerships in order to raise awareness of the issue worldwide and promote the use of energy-efficient technologies outside the EU.

These suggestions were reflected in the Commission's Communication on Energy *Energy 2020 – A Strategy for Competitive, Sustainable and Secure Energy* (European Commission 2010a), which detailed the ways in which the Commission planned to transition to a resource-efficient economy. The communication identified the need to streamline efforts to address inadequate progress in key energy policy areas such as the fragmentation of the internal energy market, insufficient security of internal energy supplies, and the disappointing nature of the member states' National Energy Efficiency Action Plans (NEEAPs). The Commission further highlighted four priority areas that, if managed and implemented correctly, could result in €1,000 of annual energy savings per European household (European Commission 2010a: 8). These included efficiency enhancements in the buildings and transport sector, industrial energy management, overall energy supplies, and an annual review mechanism for the aforementioned NEEAPs. The biggest energy-saving potential was foreseen in the buildings and transport sectors, while the main thrust of improvements in the industrial and services sector were to be made through the strict implementation of eco-design requirements for energy-intensive goods. Measures to improve transport sustainability and reduce the dependence on crude oil were described at length in an additional Commission White Paper entitled, *Roadmap to a Single European Transport Area – Towards a Competitive and Resource Efficient Transport System* (European Commission 2011c), which outlined 40 initiatives expected to lead to a 'reduction of at least 60% of GHGs by 2050 with respect to 1990' in the transport sector alone (European Commission 2011c: 3).

The Commission expected further progress on its *Energy Efficiency Plan 2011* (European Commission 2011d), which was formulated in line with the EU's target of a 20% improvement in energy efficiency and the 2020 Energy Strategy (European Commission 2010a). Therefore, it proposed a new directive on energy efficiency in June 2011 (European Commission 2011e). While the Commissioner for Climate Action, Connie Hedegaard, emphasized the directive's importance for tackling climate change (EurActiv 2011), a wave of criticism arose from environmentalists over the possibility for member states to opt out of the efficiency measures. Article 6(1) of the proposal dealt with the energy-efficiency obligation schemes and allowed the member states to set up their own schemes that should 'ensure that either all energy distributors or all retail energy sales companies operating' in the respective country's territory 'achieve annual energy savings equal to 1.5% of their energy sales, by volume' in relation to the previous year. Nevertheless, energy used in transport was excluded from the proposal

(European Commission 2011d: 20), and member states were allowed 'to take other measures to achieve energy savings among final customers' (ibid: 21).

Environmentalists were outraged. Organisations such as Friends of the Earth and the Climate Action Network argued that the directive was 'set up to fail' (EurActiv 2011) because the measures proposed were not designed to enable the EU to meet its 20% energy-efficiency target. Ultimately, the Energy Efficiency Directive, as it became known, was adopted and entered into force on 4 December 2012 (EP/Council 2012). The member states had to implement most of the provisions decided upon by 5 June 2014 and submit their National Energy Efficiency Action Plans (NEEAPs) by 30 April 2014. Despite the widespread criticism from environmentalists, the option to apply individual measures to realize energy-saving targets remained in the directive, but as consolation, the Commission expressed its intention to propose further legislation in the event that the EU failed to realize clear-cut reductions in energy consumption (EurActiv 2012).

In April 2011, the Commission proposed to overhaul existing legislation on the taxation of energy products (European Council 2003) in order to 'promote energy efficiency and consumption of more environmentally friendly products and to avoid distortions of competition in the Single Market' (European Commission 2011f, k: 1). The 2011 proposal included concrete minimum taxation levels (Table 6.2), which would have become applicable at the start of January 2013. When the proposal's text was adopted by the EP during its first reading in April 2012, and

Table 6.2 *Proposed minimum levels of energy taxation applicable from 1 January 2013*

For the purposes of motor fuels				
	CO_2 related	General consumption (€/GJ)		
Applicable:	1 January 2013 (€/t CO_2)	1 January 2013	1 January 2015	1 January 2018
Petrol	20	9.6	9.6	9.6
Gas oil	20	8.2	8.8	9.6

Kerosene	20	8.6	9.2	9.6
LPG	20	1.5	5.5	9.6
Natural gas	20	1.5	5.5	9.6

For the purposes of: agricultural, horticultural, or piscicultural works, and in forestry; stationary motors; plant and machinery used in construction, civil engineering, and public works; and vehicles intended for use off the public roadway or which have not been granted authorization for use mainly on the public roadway

	CO_2 related (€/t CO_2)	General energy consumption (€/GJ)
Gas oil	20	0.15
Kerosene	20	0.15
LPG	20	0.15
Natural gas	20	0.15

Applicable to heating fuels

	CO_2 related (€/t CO_2)	General energy consumption (€/GJ)
Gas oil	20	0.15
Heavy fuel oil	20	0.15
Kerosene	20	0.15
LPG	20	0.15
Natural gas	20	0.15
Coal and coke	20	0.15

Applicable to electricity

	General energy consumption (€/GJ)
Electricity	0.15

Source: European Commission (2011f), European Council (2003)

before it came under debate in the Council, there was good reason for optimism that the rules could be reformed. When the Council began to debate the proposal in June 2012, though, it became clear that the road to any energy tax reform would be slow at best. Noting that further work was needed on a number of issues, including, among others, minimum tax rates on products and electricity (and those used in agriculture), ETS applicable installations, and how to tax biofuels, the Council merely invited the incoming Irish Presidency to continue the debate. The Council continued to debate the legislation (in June 2013 and again in June 2014), but because directives related to energy taxation require unanimity in the Council, the new measures did not survive. The Commission withdrew its proposal in March 2015 (European Parliament 2015), and it remains to be seen whether an overhaul of the energy taxation directive will ever be adopted.

In addition to the legislation on energy taxation, the Commission supports access to other means of financing energy efficiency. Both the European Investment Bank and the European Bank for Reconstruction and Development offer relevant financial schemes, as do the Cohesion Policy Funds (2014–2020), the EU Framework Programme for Research, Innovation Horizon 2020 (2014–2020), and the European Energy Efficiency Fund.

Despite the obvious advantages of reducing energy consumption for the environment, some critics claim that the introduction of energy-efficiency standards might, paradoxically, lead to an increase in industrial emissions. They argue that such measures ultimately lead to an increase in emissions (Breakthrough Institute 2011). The logic behind their reasoning is that the use of energy-efficient appliances increases consumers' disposable income by lowering the amount and cost of energy consumed. Consequently, consumers can and do use their additional disposable income to buy other goods and services, thus boosting rather than reducing demand. The subsequent increase in the production of goods needed to meet that higher demand then leads to an increase in emissions. All things being equal, this logic helps explain why automobile use increases as efficiency standards become stricter over time. To balance this argument (and its obvious negative implications for the environment), the Commission believes that it is necessary to change its citizens' (energy) consumption habits, which in turn explains why the EU puts so much effort into public awareness campaigns.

Case study: The EU emissions trading regime

Almost a decade before the Intergovernmental Panel on Climate Change (IPCC) presented irrefutable evidence that human economic activity was directly linked to increased concentrations of greenhouse gases in the atmosphere (IPCC 2007), the EU was aspiring to lead the fight against global warming. As early as 1998, the Commission outlined unilateral steps to fully implement its obligations under the 1997 Kyoto Protocol (European Commission 1999). In 2001, after consulting a broad range of stakeholders (European Commission 2000a), it proposed to establish an emissions trading scheme (European Commission 2001a), which sparked almost two years of contentious debate over whether to make the system mandatory or voluntary, which emissions to include, and whether and how to auction allowances, finally agreeing on a voluntary auction system with the optional inclusion of other sectors and greenhouse gases (EP/Council 2003d, 2004).

Under these new rules, the EU15 sought to decrease emissions by 8% from 1990 levels by the planned second trading period (2008–2012), and member states were also required to submit National Action Plans (NAPs) identifying the necessary quantity of free allocations to be granted to national energy and industrial sectors. Unfortunately, those NAPs proved to be substantially wide of the mark, resulting in a collapse of the market in 2007. The EU tweaked the system through a series of consultations and subsequent reforms, before and after the second trading period, including replacing national registries with a single Union Registry tracking verified emissions, ownership, and purchases and sales of allowances (European Commission 2013m; see also European Commission 2010). The single registry did not solve the problem of overestimating the number of annual national allocations required and, thus, by the beginning of the third trading period, the EU market was flooded with allowances that depressed prices and reduced the effectiveness of the trading scheme.

While overestimating emissions requirements proved to be a thorn in the side of the Commission's plans to establish an efficiently functioning carbon market, the original exclusion of the transportation sector, which according to the International Energy Agency produced circa 22% of worldwide emissions in

2010 (IEA 2012: 9), proved to be an open wound. It was not as though the EU wasn't already using its internal-market powers to reel in emissions from automobiles (circa 12% of total EU emissions in 2009). It did in fact establish mandatory automobile emissions targets in 2009 (EP/Council 2009e), and would later (2013) add rules to monitor emissions from new passenger cars (European Commission 2013a). However, the real challenge was tackling air travel, foremost because of the politically contentious nature of incorporating treasured national airlines into the EU-ETS.

As far as the Commission was concerned, the aviation sector was insufficiently reducing emissions to offset increases in commercial air traffic (European Commission 2006c, IPCC 1999) and merited inclusion in the EU-ETS. Thus, beginning in 2005, the Commission initiated stakeholder consultations on when and how it could fit the aviation sector into its emissions trading scheme. The level of response, including inputs from 198 organizations and more than 5,000 individuals (European Commission 2005a), as well as the intensity of the debate that ensued, highlights the wide-ranging effect of the EU's internal-market mechanisms. Logically, European airline associations worried about their bottom line and strongly opposed the measure (AEA 2006), and member states could hardly ignore the industry's and their employees' interests. Some 3,000 enterprises operating in Europe's aviation industry employed circa 400,000 people and generated €30 billion of added value in 2006 (Eurostat 2009). By December 2006, the Commission had heard enough to formally bring aviation under the EU-ETS umbrella (European Commission 2006c). After two years of difficult negotiations, the EP and Council finally agreed to a directive (2008/101/EC) to include the sector, starting in 2012, and auction 15% of the permits (EP/Council 2008). European airlines criticized the move as unaffordable (EurActiv 2008), but they were not the only ones affected. The new rules also applied to non-European airlines flying to or from the EU. Once the CJEU subsequently rejected demands for exclusion by US airlines (Court of Justice of the European Union 2011, 2012a), on the grounds that inclusion of international aviation in the EU-ETS was compatible with international law, the EU-ETS crossed over from the internal to external energy policy dimension. This seemingly unintended turn of events resulted in the Commission temporarily deviating from the ETS Directive in November 2012 in order to seek a global agreement on the regulation of aviation emissions (European Commission 2012a).

Altogether, it took a little over a decade, but the EU developed a carbon market that covers a substantial proportion of the Union's GHG emissions, albeit a dysfunctional one that continues to be beset by oversupply. Allocated emissions' derivatives actively trade in London and Frankfurt. The glaring weakness of an oversupply of emissions allowances in the EU-ETS depressing prices and reducing some of the environmental impact hoped for by its most ardent proponents, must be balanced by the fact that the majority of EU member states successfully reduced their emissions between 2003 and 2011 (see Figure 6.7).

Finally, it must be noted that the emissions trading scheme is not the only game in town. The EU allows its industry to use a bundle of measures to reduce overall emissions, including national measures such as the Joint Implementation Mechanism (JIM) and the Clean Development Mechanism (CDM) (see, e.g. Freestone and Streck 2005). This multi-vector approach allows EU companies to avoid expensive short-term investments while concurrently strengthening the competitiveness of domestic branches, and thereby avoid negatively affecting the international competitiveness of its own emission-intensive sectors (Kreutler 2014).

The EU Emissions Trading Scheme (EU-ETS) is a demonstration of how the Commission has applied internal-market mechanisms to balance sustainability and security of supply. It directly affects the bottom line of power generators, the industrial sector, and the aviation sector. As a case, it has additional added value because its existence is central to EU plans to address climate change (the multidimensional realm) and enhance energy sustainability. By raising the cost of burning fossil fuels, the market serves a similar purpose as a carbon tax – a long sought after, but yet unfulfilled, goal of the Commission (see Chapter 3). Importantly, however, the EU-ETS is not a tax, but rather a market that provides a profit-based incentive for affected companies to reduce their carbon. In other words, heavy fossil-burning companies now pay for the right to pollute. A single allowance equals the right to emit one tonne of carbon dioxide (CO_2) or the equivalent amount of the two more dangerous GHGs, nitrous oxide (NO_2) and perfluorocarbons

Figure 6.7 *Total greenhouse gas emissions (in CO_2 equivalent) indexed to 1990*

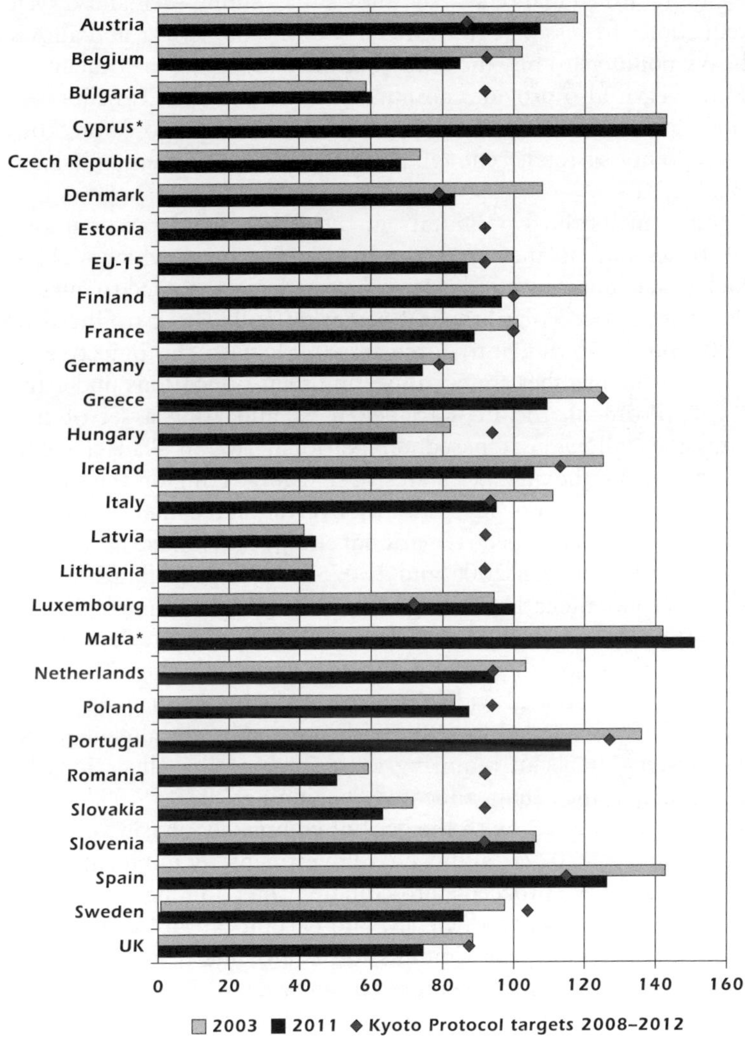

Source: Based on Eurostat (2015f) data, Council of the European Union (2002a), and UNFCCC (1998).

*No targets under the Kyoto Protocol

(PFCs). If a company uses fewer allowances than it owns, it can either save the remainder for future use or sell them to others (via well-established markets). This allows for a supply–demand-driven reduction of emissions. It awards those who modernize and allows heavy polluters to buy their way out of refurbishing their facilities. However, it also provides an important item in the EU's supranational policy toolbox, whereby heavy fines can be imposed on those who do not surrender enough allowances to cover their emissions, the EU-ETS.

Now in its third permutation, the EU-ETS has faced serious obstacles since its inception, the most important of which was the lack of harmonized standards across the Union that led to miscalculations, price volatility, and a market collapse. Introduced in 2003 (and entering into force in 2005), with the objective to support the member states in meeting their obligations under the Kyoto Protocol, the EU-ETS began by allocating a set of free emissions allowances based on National Action Plans (NAPs) submitted by the member states. Ostensibly erring on the side of caution, those NAPs substantially overestimated the allowances required by their respective national energy and industrial sectors. As a result, between 2005 and 2007, the market collapsed, with the price per allocation falling from 30 Euros per tonne to almost zero (European Environment Agency 2008: 8). Despite demanding stricter rules for the second trading period (2008–2012), the Commission received widely variant NAPs by the member states in 2008. It responded in 2009 with some substantial changes to the system. It began requiring that: some of the allocations be auctioned rather than allocated, beginning 2012, (40% of all allowances in 2013, with the goal of completely phasing out free allocation by 2027), adding a reinvestment requirement of at least 50% of the proceeds into climate protection measures, such as renewables, energy efficiency, and carbon capture and storage technologies (European Commission 2008), and revising the ETS Directive to establish a single Union-wide registry of affected companies (European Commission 2013m). Despite the fixes, however, legal uncertainties continued to challenge the Commission's authority and its legitimacy to unilaterally allocate allowances. For example, the CJEU ruled in 2009 (and rejected a subsequent appeal) that the Commission exceeded its competences when it unilaterally reduced the amount of allowances for Estonia and Poland, arguing that past emissions data did not justify the requested amount (Court of Justice of the European Union 2012).

Notwithstanding its success in reforming the EU-ETS market, the project remains a work in progress. The first phase (2005–2007) was well understood as a trial phase designed to develop the necessary infrastructure for successful trading (Ellerman and Joskow 2008), and thus the market collapse may be understood as a necessary medicine. The second phase (2008–2012) introduced some important, yet incomplete corrections. The third phase (2013–2020), currently underway, appears to be a substantially more mature market, at least in terms of volume and trading. In 2013, the EU set its Union-wide emissions' cap (plus the three EEA-EFTA states, Iceland, Liechtenstein, and Norway) at just over two billion allowances. The EU plans to further reduce emissions by annually reducing the general allowances allocated by 1.74% (based on the average of allowances issued between 2008 and 2012), or 38,264,246 per annum, thus reducing overall EU GHG output by as much as 21% from the 2005 figure by the end of the third phase (2020). As of October 2013, it was limiting emissions from more than 11,000 power stations and manufacturing plants in 31 countries, as well as aviation within and between most of them, altogether covering approximately 45% of the EU's greenhouse gas emissions (European Commission 2013n). Simply stated, the EU-ETS constitutes the world's most ambitious emissions trading scheme, and a basis from which others can learn.

Still, the real impact of the EU-ETS in reducing EU GHG production is subject to debate. While it is clear that the EU has led the way in the fight against climate change, one cannot definitively claim that the EU's emission trading scheme played a substantial role in reducing its GHG output. Other factors, such as declining rates of economic growth and milder temperatures, reduce demand for energy and heat, both major sources of emissions. Indeed, market swings can severely depress prices, and this appears to be the case with the EU-ETS. Indeed, all three phases of trading experienced sharp declines in price; or in other words, it became cheaper to pollute in each period, reducing the market's carbon-reducing effectiveness (see Figure 6.8).

Because annual allocations are determined in advance and planned across an entire phase of trading, the gap between supply and demand grew rapidly as the economy underperformed. The

Figure 6.8 *ICE futures (1-month forward) EUA in €/tonne* CO_2

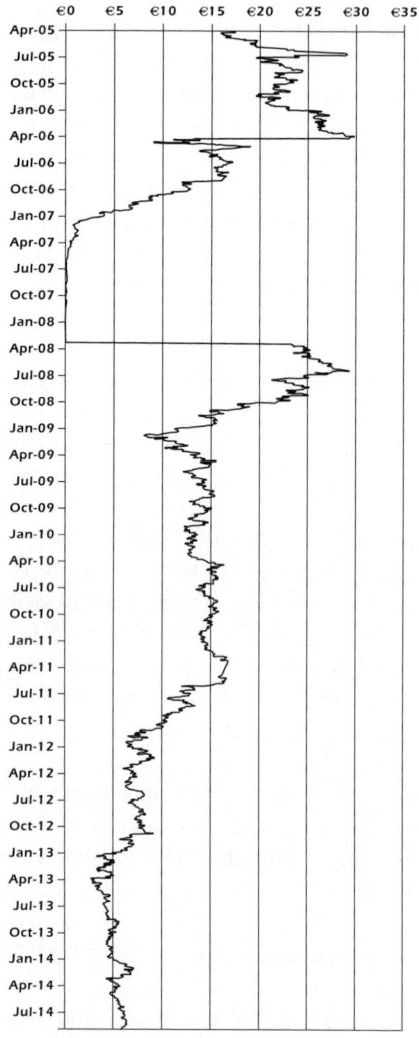

Source: Based on Bloomberg data

resulting surplus of allowances (more than 2 billion by the end of 2013; double the surplus in early 2012; and precisely the amount the Commission wanted to cut back in 2013) undermined the EU's carbon market. Recognizing its shortcomings, the Commission sought a public debate on structural reforms through a formal stakeholder consultation that ran from December 2012 through to February 2013 (European Commission 2013o). In January 2014, the Commission put forward a proposal for establishing a market-stability reserve with the start of phase four in 2021 (European Commission 2014g). Together with its initiative to postpone the auctioning of some 900 million phase-3 allowances until 2019–2020 (European Commission 2011, 2012e, 2014h), one can see how the EU was still trying to address this fundamental flaw more than a year into its carbon market's third phase of trading.

Despite its ups and downs, the evolution of the EU-ETS illustrates just how central liberal-market mechanisms are to the achievement of the other key pillars of the EU's strategy to achieve comprehensive energy security. Given the enormous effort already invested in creating its carbon market, the EU will need to implement serious structural reforms to account for the market's imbalances if it is to achieve its ambitious goal by 2030 of reducing GHG emissions by 40% from the 1990 levels.

Research and development of new technologies to mitigate climate change

Since energy was politicized in the industrial age, particularly as result of the economic and political impact of its key inventions – the coal-driven steam engine and the electric generator – government investment into new technologies and the infrastructures to support them have played a central role in driving technological change. In the last half-century, government-driven energy research and development (ERD) opened new means to producing energy (solar, wind, and tidal), extended the geographic reach of old means (offshore and deep- sea drilling), and led to ways in which fossil resources such as natural gas can be extracted from abundant shale deposits.

Europe has been a major driver of ERD for well over a century. It was the UK and Germany that moved the transport sector from coal to oil in the early twentieth century, a shift that transformed the world economy. After the 1973 oil crises, France led the way in implementing commercial nuclear technology (more than 70% of its electricity now comes from nuclear power) and today hosts the only multinational project to develop a functioning nuclear fusion generator (ITER). The US notwithstanding, Germany was among the first countries to open the market to bring small-scale, alternative producers, introducing feed-in tariffs in 2000 that required utilities to buy renewable-generated electricity (RES); and the Commission followed suit in September 2001, by adopting a directive on the promotion of electricity from renewables (EP/Council 2001).

Europe's success in driving technological change through ERD has been remarkable. However, it has also been inconsistent, ebbing and flowing over the past decades. While there was an enormous spike in such investments following the oil crises of the 1970s (almost doubling in value between 1974 and 1980 when measuring in 1974 currency values), there was an even greater decline between 1985 and 1992, and the relative level of Europe's overall ERD investments have remained below their 1975 levels ever since (Schubert 2010: 120). In the 1980s and 1990s, when energy supplies appeared abundant and Europe's economy was expanding, research into fossil fuels, nuclear energy, and even efficiency all declined sharply. Since 2003, priorities have shifted, and while investments into nuclear technologies have declined by as much as 22%, savings from other areas have led to a re-prioritization of research into hydrogen and fuel-cell technologies and renewables, including biofuels.

What all this reveals is that research and development is a core policy area in the energy domain. Recent initiatives to foster the research and development of low-carbon technologies are reflected in the European Strategic Energy Technology Plan (SET-Plan). The Communication issued by the Commission in 2007 (European Commission 2007a) identified six priority technologies: wind, solar, electricity grids, bioenergy, carbon capture and storage, and commercially viable and sustainable nuclear fusion. The EU is backing projects such as the *smart cities initiative*, targeting increased energy efficiency in Europe's cities (European Commission 2015c), and investing in the development of new large-scale turbines required to exploit offshore wind resources, as well as new materials that have the potential to substantially enhance the reliability of wind-energy technologies. This latter point is important because the bottleneck for grid integration of wind power is precisely its lack of consistency, directly resulting from periods of low wind, a fundamental problem for energy supply security (European Commission 2015d). The SET-Plan also calls for the development of advanced network technologies to make existing networks more flexible and also secure the establishment of a pan-European electricity grid that integrates the Union's diverse national networks (European Commission 2010b).

EU Research into renewable technologies is advancing rapidly. In the solar energy sector, the focus lies on the mass deployment of advanced concentrating solar power (CSP) plants, a capital-intensive technology that involves the use of mirrors to concentrate the heat of solar energy to generate electricity through steam turbines. A single CSP plant could generate enough power for almost 70,000 homes. As the name suggests, they require intense direct solar radiation and, thus, will function only in Sun Belt regions, such as the Middle East, North Africa, southern Europe, and Turkey. The focal point of bioenergy research is on the conversion of biomass (forests, waste streams, wood, and agricultural industry by-products) into fuels, heat, or electricity. Currently, research activities focus on developing advanced conversion processes and bio-refineries, with the goal to increase the share of bioenergy in the EU energy mix to 14% by 2020. European scientists expect to realize GHG emission savings of up to 60% for biofuels and bioliquids (European Commission 2015e). Concern over emissions reductions is driving the development and deployment of carbon capture and storage (CCS) systems, technologies that prevent carbon dioxide from being released into the atmosphere and, hence, are expected to play a vital role in the EU's (as well as global) efforts to reduce climate change.

It is widely agreed that the most important technological shortcoming in the energy field, and perhaps the most important for real energy security, is the lack of capability to store large quantities of electricity. Energy storage is, thus, a very important issue on the Commission's agenda, and not only in the renewables sector (European Commission 2013l, Vouldis and Vallés 2009). Major progress is being made in this area. Theorized about for years, two scientists at the University of Manchester recently succeeded in extracting graphene from graphite (Novoselov et al. 2004). (Graphene is a thin, transparent, single layer of carbon atoms, stronger than steel and highly conductive.) Their success has since initiated a wave of research that stands to revolutionize energy storage. The EU is backing the development of graphene through a €1 billion flagship project at Chalmers University of Technology in Gothenburg, Sweden, via its Horizon 2020 Future and Emerging Technologies (FET) programme, with the expressed hope that it could be used for high-power energy storage (Graphene Flagship 2015).

Finally, researchers continue to focus on how the use of nuclear energy can help realize a low-carbon economy. Nuclear-related research remains generally controversial in Europe, and became more so following the Fukushima nuclear accident in March 2011. The member states continue to be divided over what role, if any, it should play in the EU's future energy mix. For example, while several member states plan to continue using nuclear power for the foreseeable future, the German government decided to shut down all of its nuclear reactors by 2022 almost immediately following Fukushima. The Commission continues to be particularly interested in aspects related to nuclear safety of existing reactors (see Chapter 1, Section 'The nuclear portfolio') and the development of nuclear fusion power, a technology, which many perceive as a long-term energy solution that could help meet EU climate goals because it can produce large-scale electricity without causing CO_2 emissions (European Commission 2013e). In addition to supporting research into nuclear fission and radiation protection research and training (European Commission 2012f), the EU was an early supporter of the Joint European Torus (JET) project to investigate the potential of fusion power. Under the Seventh Euratom Framework Programme for Nuclear Research and Training Activities (FP7 Euratom), it earmarked almost €2 billion for fusion research, over half of which was dedicated to constructing JET's successor, the International Thermonuclear Experimental Reactor (ITER), with circa €900 million reserved for activities that included the construction of a follow-up demonstration fusion plant (DEMO) and plasma research. As a key member of the European Fusion Development Agreement (EFDA), which coordinates the European scientific and technological contributions to ITER, the Commission continues to support efforts to bring nuclear fusion online as a commercial energy source by the middle of the twenty-first century.

Concluding remarks

Since the publication of several IPCC reports on climate change, global warming caused by human activity seems to have emerged as one of the biggest threats to our society. Leading scientists draw a dark picture of the effects of man-made global warming on biodiversity, extreme weather, rising sea levels, droughts, and poverty in the world. The EU has responded to these threats by introducing ambitious targets and strategies to reduce its greenhouse gas emissions. Yet while initiatives to increase the share of renewable energy sources and develop technologies and measures to improve efficiency are widely appreciated, the environmental impact of these measures on global warming has to be evaluated cautiously.

Glover and Economides (2010: 50) posit their criticism that 'it is the erroneous *blind* faith – very different from *reasoned* faith – in the viability of renewable energy as a future energy solution that is currently driving massive public investment in alternative energy projects'. They emphasize that, instead of focusing solely on renewables, people also have to be aware of supply security problems related to these often costly and immature technologies. Dieter Helm describes the Commission's renewables directive as having 'paradoxical side effects' and questions the environmental impact of electricity production from renewables, noting, 'As more renewables are forced onto the system, the carbon price in the EU-ETS is likely to fall' and this in turn 'increases the competitiveness of coal and gas, which consequently increase their share of electricity generation, so that the resulting emissions reductions are lower and at the limit completely offset' (2012: 96). Given the slow entry of renewables and, more recently, the increased use of cheap American coal for electricity generation in the EU, Helm's analysis seems prescient.

Taking some of these concerns into account, the Commission noted in a recent Green Paper that the 'framework for 2030 must be sufficiently ambitious to ensure that the EU is on track to meet longer-term climate objectives. But it must also reflect a number of important changes that have taken place since the original framework was agreed in 2008/9'. Among other issues, the Commission explicitly referred to the 'security of supply in the internal energy market' and the 'the consequences of the on-going economic crisis' as challenges that have to be considered when talking about how to mitigate climate change (European Commission 2013f: 2). For almost two decades the Commission has strived to rein in greenhouse gas emissions, an effort characterized by bold targets and the belief that greater efficiency and more renewables could be achieved through liberalization and the realization of an internal energy market. The concomitant compatibility of competition, security of supply, and sustainability – the three pillars of comprehensive energy security – were practically dogma. It now appears that the Commission is beginning to cautiously question that narrative.

Chapter 7

External Energy Politics

The establishment of the ECSC and Euratom in 1957 marked the first modern example of a collaborative 'external' energy policy in peacetime by any state, let alone six historically warring parties. Yet despite growing to include 28 countries in little over half a century and subsequently integrating energy matters between them, developing and sticking to a common position on energy relations with states outside the EU remains one of the most divisive issues in European politics. As two observers of European politics noted, energy is both 'an issue of integration and disintegration' and one that may 'turn out to be the ultimate litmus test of [the EU's] political and economic unity' (de Jong and van der Linde 2008).

This chapter examines Europe's external energy policy in a global context and explains how it is using diplomacy and specifically rule export to ensure security of supplies from abroad with marked, but mixed success. Most work on the EU's external energy policy places emphasis on EU–Russian energy relations, the Energy Community, the role of the EU in international organizations, the building of a southern energy corridor, or the increasing role of supranational institutions (see e.g. Beyli 2012 Konstatyan 2012, Youngs 2011). Our approach differs slightly, in that we look at the EU and its energy needs and relations as a subset of the global competition for energy resources. Although the EU is not a consolidated unitary actor in the area of external energy relations, its external energy goals should 'be pursued in a spirit of solidarity among Member States' (Art. 194 TFEU), and therefore, it is helpful to understand when, where, and how this solidarity either does or does not function, and how the EU approaches trying to mitigate the latter. In so doing, we close the loop between the three dimensions of energy policy identified in Chapter 1 and show how essential, and yet distracting, the external dimension is to achieving comprehensive energy security.

The time frames required for member states to reach a consensus and implement common external policies rarely keep pace with the rapid political and market developments that occur within and between external suppliers (Schubert 2006). This is a chronic problem for the EU, whose effects manifested across multiple gas delivery shortages resulting, among others, from: disputes between Russia and Ukraine; from shortfalls in LNG deliveries due to a spike in demand from Japan following the Fukushima disaster (Japan relies on LNG imports for virtually all of its natural gas supply and pays a premium, e.g. 35–50% above average EU prices over the first quarter of 2013); or as cheap coal flows to Europe from a US surplus due to the latter's internal success in producing unconventional natural gas. The fact is that the EU faces stiff competition from other large consumer countries on the one hand and is subject to market forces beyond its control on the other, both of which affect the availability and pricing of key fossil fuel resources.

At the core of the EU's external energy strategy is the need to secure the stable inflow of the resources it needs. In an international context, those needs translate into long-term reliable access to oil, natural gas, coal, and uranium. Of the four, it is the struggle to secure reliable access to and transport of oil and natural gas that have the most substantial effect on Europe's long-term energy security. A *Public Consultation Document* from the Commission notes that the 'EU energy system is not isolated' and '[f]or the next decades, access to hydrocarbon resources will remain of strategic importance' (European Commission 2011h). The EU imported over €400 billion in oil and gas in 2012, which is more than double the average annual import cost between 1990 and 2011 (European Commission 2014) and indicative of its vulnerability to supply and price shocks.

In its 2011 Communication, *The EU Energy Policy: Engaging with Partners Beyond Our Borders*, the Commission identified four objectives to ensure its energy supply (European Commission 2011g). First, it wanted to improve internal coordination by establishing an information exchange mechanism on agreements between member states and their suppliers. The Commission argued that this would strengthen the external dimension of the EU's internal energy market and allow it to make agreements with third-party countries, including Russia, at the EU level. Second, the Commission wanted to strengthen its external energy supplier partnerships by establishing standardized rules for EU relations with suppliers and transit countries, including legally binding nuclear safety standards – what may also be called rule export. Third, it wanted to mainstream energy in its development policy. Fourth, it wanted to rank its energy partners, and thus treat them accordingly with appropriate instruments, by proposing to set up a Strategic Group for International Energy Cooperation in order to centralize discussions and help foster solidarity. The first two of these four objectives are real policies with measurable outcomes that could affect EU energy supply security. It is difficult to identify an immediate impact of the latter two, however. The fact is that EU external energy policy is best explained in terms of its dependence on and relationships with foreign suppliers, and particularly, the need to reduce or dissipate the former and improve the latter.

This chapter begins with a brief overview that sets the context within which the external dimension for energy operates, including the core issues and obstacles it faces. Most notable among them is the need to achieve supply diversity and reliability and put solidarity into practice in light of an incomplete internal energy market. We then look a little closer at the relationship between sovereignty, solidarity, and energy resources before describing the EU's most important external partners, venues, and engagements, in particular the Energy Community and the EU's special bilateral energy relationships. We then turn to the thorny politics of pipelines and the maturing LNG options, examining the infrastructure in place and its consequences for policy. We round off the chapter with an examination of four geopolitical challenges that will affect, if not shape, the continued evolution of EU external energy policy.

Setting the context

The following subsection sets the context for understanding the EU's external energy policy and, based on the premise that there is a need to acquire additional resources from abroad, explains why the Union's external energy dimension is fraught with economic, geopolitical, and even ethical risks.

The international energy market is imperfect. It is subject to speculation, political interests, and machinations, which add volatility to prices and, on occasion, disrupt the smooth flow of supply to consumers. Therefore, Europe's common external energy policy needs to be understood in geopolitical terms. International coal, uranium, and oil markets are relatively stable and globalized. Oil prices tend to be more volatile, but oil has not been subject to major supply disruptions since the late 1970s. Although natural gas is increasingly being sold and transported via ship (31% of all traded gas worldwide and 26% of EU imports in 2012), the EU still receives the vast majority (76%) via pipeline (BP 2013, Ferrier 2013). Even if the LNG market matures such that supplies are sufficient to decouple the price of gas from oil – a circumstance that the EU is trying to help bring about by liberalizing its gas market using regional hubs – the sheer volume of gas consumed in Europe predicates that the EU will continue to depend on pipelines, and the suppliers and transit states connected to them, for the foreseeable future.

Acquiring energy resources of any type entails complications. Renewables such as wind and solar energy are widely distributed, but costs and capabilities vary widely based on geography and climate. Fossil fuels are geographically concentrated, environmentally messy, but they also are fixed and reliable. Coal and uranium, both used for electricity generation, can easily be transported in large quantities by ship or rail. Oil is shipped by tanker or, when geographic proximity permits, by pipeline. As there are many different suppliers of petroleum, the international market for oil is highly competitive and prices are market driven.

Until the recent advent of technologies allowing gas to be liquefied and shipped from distant locations, natural gas was subject to regional politics. Shipping gas via pipeline from source to destination inexorably links specific suppliers to specific markets based on long-term contracts and fixed prices. Consequently, countries heavily dependent on imported natural gas tend to be geographically and politically limited in their options to find alternative suppliers. Nowhere is this more evident than in Europe.

Indigenous reserves, the geographic distribution of external sources of gas and oil, technology, and the willingness of the EU's competitors to cooperate are all factors that affect the EU's ability to secure its foreign fossil fuel resources. Geography colours the EU's strategic energy interests and options. While EU members Denmark, the Netherlands, Germany, and the United Kingdom together sit upon around 2.2 trillion cubic meters (tcm) of proven natural gas reserves (BP 2013), those reserves amount to only 4% of the world's total. At current production and consumption ratios, if the four were required to provide the Union with all its needs, they would exhaust those reserves in less than a decade. Europe simply has no choice. It must look outside its borders for natural gas. Three factors determine Europe's ability to import those resources: the international reach of its pipeline network, the quantity and capacity of its LNG import terminals, and the competition it faces for the sources to feed those pipelines and ships.

To meet these needs the EU looks to Russia, the Middle East, the Mediterranean Basin, and places as distant as Central Asia and as difficult to operate in as the Arctic. Europe's nearest gas suppliers are aligned along its frontiers: Norway is to the north-west, Algeria across the Mediterranean, and Russia to the east. Plans are afoot to create a fourth corridor to the south-east to acquire Caspian and eventually Iraqi, and even Iranian, resources, both of which are constrained by political rather than technical issues. Deposits in the Arctic are still undeveloped and their dimensions remain largely uncertain, as the very territory itself and resources it holds are in dispute. The world's remaining natural gas reserves are distributed across the globe, in some cases in large quantities, and slowly becoming available as LNG.

The EU is a unique energy importer because it oscillates between decades-old attempts to agree on a common external energy policy on one hand and national interests and egoism on the other, the result of which is a series of overlapping arrangements to import supplies. That redundancy actually acts as a positive factor for EU security of supply. However, the lack of unity hinders the realization of a single internal energy market and creates economic advantages and disadvantages, diminishing Union-wide returns on any oversupply. It also shapes the external relations of individual member states. Austria, for example, imports natural gas from Russia via Ukraine (4.7 billion cubic meters in 2012), and then exports the vast majority westward, for a profit, benefiting its neighbours, but at the same time tying its economic fortunes to its relationship with Russia. Austria started importing Russian gas in 1968, long before it joined the EU, yet its case is emblematic of the energy divide found within Europe, one shaped by geographic proximity to land-based suppliers or maritime access. Coordinated and used for mutual benefit, these internal divisions can actually enhance the EU's security of supply, but left unused or, worse, misused, it can also be the source of internal disputes over both external policy and internal solidarity.

The EU's internal divisions demonstrate the importance of geography in understanding the larger international energy game. The resources Europe requires are not evenly distributed around the globe. Australia and Canada are the world's primary suppliers of uranium ore. Coal is available in large quantities in the US, the EU, Russia, India, Indonesia, and China. Most of the world's proven oil reserves, however, are located in the Middle East (with vast reserves of unconventional oil found in the US, Canada, and Russia); and most of the planet's natural gas comes from Russia, Qatar, and Iran, and increasingly from the US, with vast, but yet untapped, quantities in China.

Europe is highly dependent on foreign suppliers for each of the primary resources. Uranium ore and coal, both used for power generation, do not pose nearly as many problems as oil and gas, because the suppliers of these resources are stable and reliable, as are the means to ship the minerals to and within the EU market. Nevertheless, the EU's commitment to reduce greenhouse gas emissions, and its earlier shift away from oil for use in power generation in response to the 1973 Arab oil embargo, only increased its dependence on its less environmentally damaging substitute, natural gas. Oil, meanwhile, remains essential for the transport industry. EU dependence on foreign-supplied oil and natural gas in 2012 was a staggering 87.4% and 65.3%, respectively, in 2013 (Eurostat 2015). It was also concentrated. Russia alone accounted for more than a third of the EU's oil imports, and Russia, Norway, and Algeria combined to supply 71% of its gas imports. According to ENTSOG, while 'no Zone had a supply dependency on Norwegian, Algerian, Libyan or Azeri supplies over 20% [...] a range of Zones [were] reliant on Russian gas with 10 Zones having a supply dependency of 60% or more in 2013' (ENTSOG 2013a: 9).

Yet as bleak as this picture may appear when it comes to oil, the EU has actually diversified its gas suppliers since 2006, when the same three suppliers accounted for almost 85% of EU gas imports. This is in part due to the economic downturn that began in 2007 and growth in LNG deliveries to EU terminals, which in 2011 accounted for 19% of its gas imports (Ratner et al. 2013). Also, because the international market to buy and ship oil is dynamic and globalized, sudden and extreme oil shortages are few and far between. Price volatility is part of the market, but as long as global production capacity exceeds demand, short- and medium-term access to supplies remains relatively secure. To hedge the risks of significant or prolonged supply disruptions, member states also maintain a minimal amount of strategic oil reserves that, in theory, should cover up to 100 days' worth of missing imports. Although no similar arrangement formally exists with natural gas, gas hubs and storage facilities do exist, and in early 2014, national inventories of gas equalled between 60 and 90 days' worth of demand (Gloystein and Kahn 2014; European Commission (2009e)).

Awareness of the risks to its gas supplies inspired the member states to increase their cooperation, and there has been good progress since 2004. For example, the EU established a new legal framework (Directive 2004/67/EC) to safeguard gas supplies (EP/Council 2004a) and created two new high-level bodies, the EU Network of Energy Security Correspondents (energy experts from the member states) and the Gas Coordination Group. Following the gas supply crises of 2006 and 2009, the EU passed another regulation (994/2010) on the security of gas supplies (EP/Council 2010a), which aimed at harmonizing standards and setting out rules for crisis planning, as well as a coordinated response, in the event of future supply disruptions. Building upon the general principle that member states should diversify routes and sources of supply, the regulation called upon the member states to ensure that 'transmission system operators shall enable permanent bi-directional capacity on all cross-border interconnections' between the member states unless they request an exemption or make an alternative proposal for bi-directional capacity concerning the reverse direction (European Parliament and Council 2010a: Articles 6.5 and 7). It set a 3 December 2013 deadline to 'adapt the functioning of the transmission systems in part or as a whole' in order to enable those reverse flows. To meet these new requirements, the regulation also stipulates that the member states at least have to consider increasing their storage capacities. However, it was not made a requirement as in the case of oil. As of early 2014, EU law still did not require member states to maintain strategic reserves of natural gas. Thus, during emergency shortages, existing or ad-hoc arrangements have to be quickly implemented, making use of the member states' autonomous potential to store and move gas between them.

Complicating matters, Russia delivers most of its gas supplies to Europe via a pipeline network built in the days of the Soviet Union. Designed originally to interlink the now defunct COMECON economic block, with the added value some years later of getting hard currency from Western Europe through exports, those pipelines cross independent states to reach the borders of the EU, adding another type of player to the calculation, the energy-transit state. Disputes between Russia and Ukraine, the latter being the primary export corridor to Europe, have led to significant disruptions to EU supply and complicate solidarity-based responses to Russian intransigence in multilateral organizations or its security activities in the Caucusus. The most significant crisis occurred in 2009 when, between 7 and 20 January, Russian gas supplies to Europe were completely shut off. While some EU member states could rely on other sources as well as reroute some of their own surpluses eastward, Slovakia, Poland, Hungary, and Romania were suddenly cut off in the cold of winter, a circumstance which Europe has endeavoured to avoid in the future (European Commission 2009).

The recurrent Russia–Ukraine energy dilemma is indicative of the risks of energy interdependence, and in particular the politically sensitive issue of pipeline politics. Pipeline politics revolve around the routes pipelines take from source to destination. It is not merely a matter of supply. Since pipelines cross borders, national sovereignty is affected, and construction requires large amounts of money and long-term planning. While suppliers and consumers both want to maximize direct access, minimizing costs for consumers and maximizing profits for suppliers, transit states through which the pipelines flow take cash fees or i- kind supplies, raising costs and reducing profits. When disruptions occur along a transit route, everyone loses. The choice of routes affects national incomes and consumer supply security. As a result, international competition over which lines to build, which routes to follow, and which volumes of supplies to deliver are often the source of heated negotiations (at both the government and commercial levels), and have provided the basis for bilateral cooperation and competition. Such was the case when Russia and Germany agreed to jointly build Nord Stream, a project that, despite original opposition, has proven to be a positive addition to Germany's security of supply.

In addition to physical pipelines and their routes, the EU also needs to consider a complex set of economic, technological, and political issues that directly or indirectly affect its security of supply of gas and oil. Any external energy strategy involves investments in new energy technologies, the construction of deep-water ports, upstream exploration and production, downstream distribution, environmental considerations, competition for supplies, and common regulatory frameworks. Equally important is the economic and political conditions of its major suppliers. Many of them (Algeria, Norway, and Russia in particular) rely on energy exports to the EU for substantial shares of their GDP and, with the notable exception of Norway, they tend to lack transparency, and suffer from corruption, domestic inequality, and political instability (Schubert 2006). Consequently, the EU not only faces material risks associated with geographical access, it also faces legal and ethical dilemmas over governance standards, transparency, and economic development in each choice of supplier. For the EU and its energy companies that means guaranteeing the security of its foreign direct investments in the upstream sectors of key energy partners such as Russia and prospective ones like Azerbaijan. To that end, the EU employs traditional economic statecraft at the bilateral and multilateral level. Through the former it has secured supplies from Azerbaijan and established better communications and information exchanges with Russia; through the latter it has worked to export its internal energy market rules, primarily through the Energy Community.

Energy geopolitics also includes struggles for regional domination and spheres of influence. The EU's struggle to diversify its sources and suppliers place it right in the middle of the some of the world's longest running and thorniest venues of international competition. For example, not only does the Caspian region possess oil and natural gas in abundance, its position sandwiched between powerful neighbours Russia, Iran, and China makes it the perennial target of political manoeuvring. The US is no less involved, as it sees Central Asian independence as a bulwark against Russian expansion and Iranian influence. Early twenty-first century activity in the region is strikingly similar to The Great Game of the nineteenth century, when the UK and Russia drew up the region's borders. Access to African energy resources is no different, as Chinese, Russian, American, Indian, and European companies compete fiercely to secure deliveries to their respective domestic markets, generate revenues, and control physical access to resources headed to other markets. Moreover, because many of Europe's competitors operate state-owned energy conglomerates, the quest for energy resources is not a mere matter of market economics, but rather a geopolitical game of Monopoly.

One of the large global players, the United States, broadly shares the EU's international energy goals, particularly as pertains to establishing universal norms and laws, and soon may develop into a natural gas supplier (EIA 2014, IEA 2013). China is a powerful competitive consumer also actively seeking to secure resources abroad. It is doing so by integrating its state-owned companies into the global supply chain, of which the likely outcome is an addition to rather than a drawdown of supplies in the market, because the distance between mainland China and its companies' remote activities makes exclusive deliveries too expensive. However, Russian investments appear to be less oriented towards adding supplies to the global market than at maximizing Moscow's share of existing global trade, an approach that meshes with Vladimir Putin's notion of a Russian renaissance.

What is particularly troubling for the EU is that both Russia and China prefer a multipolar rather than non-polar world, where regionalization trumps multilateralism and energy is a central component of their respective strategies (Die Welt 2008, Isachenkov 2003). A multipolar world without a commensurate increase in global norms and rules would not bode well for advocates of global governance, and could easily endanger institutions such as the Energy Community. Russia's hard-handed tactics in Georgia and Ukraine may be an indication of what is to come. Meanwhile, China singled out energy as a fundamental component of its security environment (PRC 2006, 2009: 4).

Instability and uncertainty about North Africa's regimes also plays a role in shaping the EU's external energy planning. Political unrest is prevalent across the Maghreb. If and when violence erupts, key components of national energy infrastructures could be disrupted or destroyed. In Algeria, this could affect gas supplies to Spain and France. In Egypt it could reduce or even halt shipping traffic through the Suez Canal, a waterway of tremendous value to European energy security. Post-war Libyan unrest has already led to a several month-long halt in gas exports to Italy in 2013. Added to the political uncertainty is the economic one, brought on by the region's future domestic demand, which is expected to eat into energy exports (Westphal 2011).

Prior to the oil crises of the 1970s, privately owned international oil companies (IOC) controlled most of the world's oil reserves. Today, private, commercial energy companies control very little of the world's physical oil and natural gas resources and have to compete with state-owned energy companies from Russia, China, and India, among others, for access to resources, ownership of pipelines, ports, refineries, and distribution facilities. It is almost impossible to separate the actions of Russia's Gazprom or China's CNOOC from the foreign policies of their respective governments. All of these factors, from political instability to the increased role of quasi-state commercial enterprises, from the potential for new discoveries to structural changes in the international energy market, means that the EU must diligently pursue a multilateral legal framework for the exploration and delivery of fossil fuel resources.

Finally, similar underlying motivations drive both the EU's external energy policy and those of its suppliers. Whereas the EU focuses on security of supply – reliable delivery from multiple suppliers with 'transparent and non-discriminatory security of supply policies' (Council 2004) – its suppliers, seek security of sales – reliable, long-term commitments by multiple customers purchasing on long-term contracts, and investing in upstream exploration and infrastructure with as little market regulation and meddling as possible. Thus, security and diversity are equally important to planning for both parties. They simply come with different priorities. On the price side, exporters want prices just high enough to stabilize their income and, often, their national budgets, but low enough to offset downward pressure in demand. Buyer diversity translates into greater consumer competition for resources, adds upward pressure on demand, and increases revenues, which in turn drives a deep wedge between EU member states as well as between the EU and its primary competitors.

In summary, the global energy game is one of long-term investments and strategic positioning. Poor planning or a lack of commitment can leave countries vulnerable and dependent upon the vagaries of international politics. Consequently, every EU energy initiative on the global stage is under relentless stress. These conditions help explain why the EU focuses its external energy policy on integrating suppliers, consumers, national interests, and geopolitical forces in regional, and ultimately global, multilateral legal frameworks designed to mitigate political obstacles and maximize the market benefits for its participants.

Sovereignty, solidarity, and energy resources

Centrifugal forces hamper the EU's pursuit of a common external energy policy. Issues such as sovereignty, trade relations, foreign dependency levels, the absence of a global legal framework for the exploration and trade of energy resources, and the anarchic nature of the international political system all combine to weaken European solidarity. Consequently, the EU's outward approach towards energy more often resembles a set of independent islands than it does a unified block. Of all the forces that test EU solidarity, national sovereignty is the most powerful, and energy is one of the issues at its core. Vital to economic growth and national security, governments loath relinquishing such powers as the control of mining rights, transit permissions, customs and border controls, protection of critical infrastructure, and even market regulation. The member states have transferred some energy competences to the EU's supranational institutions, particularly as pertains to market oversight. However, because the member states retain full autonomy over their respective national energy mixes, each state also maintains independent relationships with foreign suppliers in the form of bilateral agreements, in many cases taking the form of long-term supply contracts. These competing foreign relations exacerbate existing internal divisions over external policies and hamper the Commission's efforts to formulate a unified external approach to energy.

Member states vary significantly in size and level of dependence on particular energy supplies (see also Chapter 1 on the EU's energy portfolio). Natural gas provides a case in point. The EU's largest energy markets are located in Western Europe. France and Italy import large amounts of natural gas from multiple external sources, largely because of LNG and proximity to North Africa. In both cases, however, natural gas does not constitute the largest share of their primary energy needs. The same is true of Germany, which imports from Norway, Russia, and the Netherlands (to a lesser extent also Denmark and the UK). The opposite is the case in many of the Eastern and Baltic member states where the economies are significantly smaller, but foreign dependence on a single supplier is dangerously high (in some cases approaching 100%). Moreover, some states are more dependent on natural gas for their electricity generation than others. Those that currently rely on coal to generate their electricity are under pressure to increase their power-generating share of natural gas or renewables to meet emissions targets. Variances in Europe's domestic energy mixes and the incomplete nature of the internal energy market all but guarantee fundamentally different priorities for each member state. Thus, the very forces that should unite the member states actually divide them.

While matters of sovereignty and trade dependence pull member states away from each other, changes in the international energy environment over the last four decades considerably complicated matters. In 1972, one year before an Arab oil embargo sent

European countries dashing for alternatives, OECD Europe generated more than a quarter of its electricity with oil (IEA 1997). Following the embargo, Western European states and the US moved to remedy their dangerous dependency, banding together to establish the International Energy Agency in an attempt to coordinate the demand side of the market, and began switching fuels and building emergency stocks. Most increased their use of coal and some turned permanently towards nuclear power. One example of the latter was France, where oil accounted for as much as 49.3% of electricity generation in 1972 (Daintith and Hanch 1986), whereas by 2006, oil accounted for a mere 1% of French electricity generation. Yet, France's shift to nuclear was not uniform across the EU.

The risks associated with oil dependency, combined with the existential danger of nuclear accidents, impelled many EU countries, including those that would join later, to turn to natural gas. This had the positive effect of diminishing the European electricity market's dependence on Middle East oil. However, most EU members ended up exchanging foreign dependence on one source for another. Large discoveries in the North Sea provided a reliable alternative at first, but as production peaked and demand increased, Russian exports grew in importance. The EU's slide into external dependence for natural gas accelerated in the late twentieth century, as environmental concerns and fears of global warming pushed Europe to increase its share of natural gas in power generation as a means to reduce harmful emissions. By 2011, oil's share of EU electricity generation had declined to almost 2%, while natural gas had risen above 22% and constituted 35% of gross inland consumption (European Commission 2013b). Still, it would be difficult to argue that the EU is substantially more secure today than in the past, and its energy crises keep returning like clockwork. If the 1970s was defined by Europe's dependence on Middle Eastern oil, the early twenty-first century is defined by its dependence on Russian gas.

Much focus is placed on the security of supply, but it is actually the insecurity about existing supplies and specific suppliers from abroad that is the driving force behind the pursuit of a common energy policy. Although gas and oil are very different products in terms of their downstream delivery and consumption, past experiences involving oil are useful in understanding the importance of gas. The history of oil teaches us that international security conditions can change rapidly, and more often than not, the EU will have little to no ability to prevent serious disruptions. This was true in 1956 during the Suez Crisis, the 1970s with the Arab oil embargo, the 1980s' tanker wars in the Persian Gulf, and more recently with Russia over Ukraine (gas) and sanctions on Iran have kept vast gas resources that could flow to Europe buried underground.

The EU must diversify not only the mix of fuels it uses but also the suppliers and transit routes in order to maximize the security of its external energy supplies. While integration of member state networks is moving forward, the task remains incomplete. This all but forces individual states to coordinate their own imports bilaterally with the nearest or most accessible suppliers. Along its eastern frontier, this translates into relations with Russia. Along its southwestern stretch, this means Algeria, and to the northwest, Norway. Therefore, competing EU energy interests divide along geographic lines according to respective dependence, another fact that divides Europe rather than unites it.

Europe's past energy crises demonstrate the need for emergency planning. Maintaining multiple suppliers and routes, a large domestic storage capacity, fuel-switching plans, a high-capacity infrastructure, and being part of an internal market are each essential elements of such planning. The first two (suppliers and routes) are external in nature. The remaining four (stockpiles, fuel-switching plans, infrastructure, and the creation of a fully unified internal energy market) are matters of internal EU politics. None of these issues exist in a vacuum. For Europe to formulate and execute a coherent external energy policy, it needs to resolve its internal issues.

Externally, the EU is expanding the outer boundaries of its internal market through diplomacy and rule export. It uses multilateral agreements and programmes such as the Energy Charter, Black Sea Synergy, and Euro-Mediterranean Cooperation, as well as institutions such as the Energy Community, the WTO, and the International Energy Agency to pursue a common set of standards for protecting investments and deliveries. It is trying to integrate the national energy grids and economies of its neighbours and suppliers directly into the European system of governance. This is particularly germane in the context of pipelines, where transit agreements, fee structures, ownership rights, and transparency issues all play a vital role in Europe's energy security.

The EU's strategy has been largely successful along its periphery, but it is not free of obstacles. While EU influence is strongest along its borders, it is uneven. For example, European companies and politics dominate North African energy production, but do not do so in Russia, the Caucasus, or Caspian Littoral. Algeria criticizes Europe's strategy of demanding transparency, open access, full ownership, and market liberalization in exchange for aid as amounting to unacceptable meddling (Schubert 2007). Pushing Turkey to accept a role as transit state (as opposed to importer/exporter) challenges Ankara's cherished role as arbitrator between East and West, not to mention cutting away the country's profits. Russia sees the Caucusus within its sphere of influence and considers control of its own mineral resources as a matter of national pride and security and, thus, objects to EU efforts to push for open access into its domestic energy market or make independent energy agreements with Ukraine. The EU has had some success in Central Eurasia, but its efforts also inspired Russia to propose an alternative (first the South Stream pipeline, and when that failed, then the Turkish Stream) in order to secure its primary role as the EU's premier supplier of natural gas.

The circumstances with Russia are special and unique along the EU's periphery. Neither Norway nor the countries of North Africa are dominated by one foreign power, as is the case of those countries that were part of the former Soviet Union. EU attempts to influence affairs in the Caucasus were set back in 2008 when Russia invaded Georgia, and again in 2014 when unrest in Ukraine led to a not-so subtle military intervention and annexation of the Crimea by Moscow. Disruptions along Russia's East European gas and oil supply routes through Ukraine and Belarus demonstrate Europe's weakness in responding to Russia's drive for regional hegemony.

Each of these examples illustrates a poignant problem for the EU, namely that pursuant to its external energy needs, the EU finds itself marching into geopolitical challenges rife with troubles, four examples of which we provide towards the end of this chapter.

External actions: Europeanization, diplomacy, and rule export

EU external energy policy matured substantially in the early twenty-first century. Supranational institutions are increasingly taking the initiative in the field of energy, particularly as pertains to setting targets. The EU is even supplanting the member states as the principal initiator of external energy policies (Beyli 2012). For example, in 2011 the Commission published a Communication on security of energy supply and international cooperation – *The EU Energy Policy: Engaging with Partners Beyond Our Borders* – where it laid out four main priorities 'to further develop an external energy policy' (European Commission 2011g), and in November 2011, the Council approved a similar set of priorities (Council 2011). In brief, the common priorities suggested by the Commission and set by the Council were to: (1) strengthen internal coordination of external energy policy, (2) strengthen EU cooperation with third-party countries, (3) deepen energy partnerships, and (4) support developing economies.

To strengthen coordination, the Commission and Council set out to institutionalize an information exchange of bilateral agreements that member states either already have or negotiate with third-party countries, and to create the possibility for the EU to directly negotiate with third parties, particularly in the area of strategic infrastructure projects. They also proposed to increase both the number and quality of meetings of the Energy Council in order to prepare a consistent and coordinated message for all high-level meetings of international organizations. To strengthen EU cooperation with third-party countries, they agreed to utilize multilateral instruments to extend the application of the EU's regulatory framework by enlarging the 'the Energy Community Treaty to neighbouring countries' (Council 2011: 4), including Turkey and 'faster integrating Ukraine into the Energy Community' (European Commission 2011g: 5). They also identified the Southern Corridor, the Eastern Corridor, and the Mediterranean Basin as 'priority areas for energy projects', and recognized the dual needs to extend the existing early warning mechanisms (EWM) for possible disruptions or dips in supply, and to expand cooperation in research and development of future-oriented energy technologies. To strengthen its energy partnerships, the member states agreed to expand where possible and establish where necessary energy partnerships with North Africa, the Middle East, the Caspian, and the Black Sea; and committed to resolve outstanding issues with Russia under the framework of its Partnership and Cooperation Agreement (PCA), including access to energy resources and infrastructure, investment protection, and non-discriminatory pricing of energy resources. Finally, both the Commission and the Council documents promote the idea of supporting developing economies through a basic set of principles such as reciprocity, transparency, investment protection, and sustainable development of third-party energy sectors. They argue that doing so will diversify suppliers in the market and, by tying EU cooperation in the energy sector to institutional and legal reform, help contain the negative environmental consequences that the requisite increase in exploration and production will generate. Together, these actions represent the bulwark of contemporary EU external energy policy.

The Commission spearheads most external energy policy actions with heads of government playing their part as well. Although its role remains quite limited, the EU's External Action Service is increasingly getting involved in some of the second-level discussions and dialogues (Van Vooren 2012). In May 2013, the European Council instructed the Council to review developments regarding EU external energy policy, and in December, the Council issued a follow-up report that reaffirmed the 2011 Council conclusion as constituting the first comprehensive EU external energy policy (Council 2013).

As a result, the Commission and the member states increased both the quantity and quality of cooperation on subjects relating to external energy matters. Energy became a regular topic of the Strategic Group for International Energy Cooperation (SGIEC), which had already put on its agenda energy relations with China, Ukraine, North Africa, the US, and Russia. The SGIEC, established in 2012 on the recommendation of the Commission in its Communication on security of energy supply and international cooperation (European Commission 2011g), brings together representatives from the member states and 'relevant EU services' to ensure better coordination of the EU external policy issues and strengthen the negotiating position of the EU in international fora by developing a common message on EU cooperation with third-party countries. Commission representatives now raise energy issues at virtually every international forum and integrate the topic into more Council formations, including those not related to energy. As a result, there appears to be a better understanding of common principles and priorities among the member states.

In fact, since 2011, the Commission has made significant progress on several external energy fronts. It signed Association Agreements with Moldova and Georgia, both of which contain key energy provisions. It progressed in negotiating an agreement on electricity networks and market interfaces between the Baltic member states and Russia and Belarus (Council 2013). It initiated adding a Protocol on electricity to the EU–Switzerland Free Trade Agreement, and negotiated a legal framework for a trans-Caspian natural gas pipeline system with Turkmenistan and Azerbaijan. It also concluded a Memorandum of Understanding on energy cooperation with Algeria, and, in December 2013, initiated talks on forming an Energy Community with the members of the Union for the Mediterranean (UfM). In terms of key partnerships, the Commission solidified its working relationships with the US, China, and OPEC, and engaged Russia through the EU–Russia Permanent Partnership Council (PPC) on Energy and the EU-Russia Energy Dialogue. An outgrowth of the latter was the March 2013 'EU–Russia 2050 Roadmap', in which the Commission set itself the lofty goal to bring Russia and the EU together in 'a common, subcontinent wide, energy market' by 2050. The Roadmap is, however, non-binding. Given that it will 'require the gradual approximation of rules, standards and markets in the field of energy', such goals may prove overly optimistic (see European Commission and Russian Federation 2013).

Perhaps the most visible sign of progress towards a common external energy policy was getting approval for the controversial proposal to set up the exchange mechanism on intergovernmental energy agreements. When initially proposed in 2011, member states were sceptical. It seemed like a direct challenge to their sovereignty, a matter of competence reserved for the member states. However, as a sign of the growing solidarity – at least in terms of coordinating external supplies – the Council and Parliament passed Decision 994/2012/EU obliging member states to submit their already existing legally binding agreements that have 'an impact on the operation or the functioning of the internal energy market or on the security of energy supply and their new agreements, once ratified' (European Commission 2013g). The Commission now plays an oversight role in the bilateral energy relations of the member states, controlling whether their agreements adhere to internal market rules. It can now not only provide assistance during bilateral negotiations but also order ex ante revisions. By the conclusion of 2013, the Commission had already evaluated more than a hundred separate agreements.

All of these developments point to clear progress in the external energy domain. The EU in 2014 is much closer to speaking with one voice than it was in 2010. However, solidarity is a goal and a process, not an irreversible fact or irrevocable principle. The fact remains that there is much more still to do than has already been done. For example, in terms of speaking with one voice, the EU needs to increase the consistency and coordination of the messages its representatives deliver and the outcomes its negotiators pursue in dealings with international organizations. In its review of the Commission's progress, the Council argued that this can be achieved by preparing what it calls 'lines to take' (Council 2013). Still, establishing a common voice is not a replacement for shared interests or principles. Coordinating, and even consolidating the latter two, requires significantly enhanced interaction at the EU level between the relevant authorities and representatives of the member states. To an extent this has already been achieved, but it is incomplete. The Strategic Group for International Energy Cooperation, for example, met only six times in total between 2012 and 2013, and although the Transport, Telecommunications and Energy Council (TTE) meets on average six times a year, external energy issues are not always on the agenda, which is an easy opportunity lost.

Despite progress on the Caspian front, there is an urgent need to further develop strategic energy corridors and, in particular, to fund key projects of common interest (PCI), such as the Trans-Anatolian Pipeline (TANAP), that open new access routes to gas and oil. This latter point takes on special meaning when one considers the effort and political weight thrown behind Nabucco, a project that, despite Brussels' support, ultimately had to be truncated to a line entirely within the boundaries of the EU. Equally important, the EU needs to find ways to capitalize on new discoveries, such as those in the Levant Basin and the Black Sea, either through financing, expanding the Energy Community southward, or creating a Mediterranean Energy Community. The latter option is problematic because Israel and Lebanon have no official relations and disagree over who has rights to develop newly discovered offshore fields. On the broader regional scale, the EU needs to continue to expand the Energy Community's reach and depth, use its Neighbourhood Policy in general, and the Eastern Partnership (EaP) (European Council 2009) and the Union for the Mediterranean specifically, to extend the coverage of the EU's energy acquis, as well as make certain that any final arrangement it reaches with the US on transatlantic trade, the so-called Transatlantic Trade and Investment Partnership (TTIP), includes an energy chapter.

Main partners, venues, and engagements

Diplomacy and economic statecraft are the primary tools through which the EU secures its external energy interests. Those interests translate into actively trying either to export or extend the coverage of its internal energy rules through bilateral agreements, multilateral institutions, and international fora, such as the Eastern Partnership and the Union for the Mediterranean. The same is true with its 'energy dialogues', which it maintains with Russia (since 2000), Brazil (since 2007), China (2005), Japan (2007) and OPEC (since 2004), and through its bilateral relations with important energy partners such as Algeria, Iraq, Turkey, Ukraine, and the US, the latter of which established a ministerial-level Energy Council in 2009. The Presidents of the European Council and the Commission, as well as the DG Energy, frequently take part in bilateral summits. Furthermore, EU representatives of various levels work regularly with counterparts in major international venues including the IEA, IAEA, the International Renewable Energy Agency (IRENA), the ECT, G8 and G20 summits, the International Energy Forum (IEF), IPEEC, and the UN; they even attend ministerial meetings at OPEC, and regularly communicate with representatives of the Gulf Cooperation Council (GCC) through the EU–GCC Energy Experts Group. In 2013, such efforts with the IAEA led to a Memorandum of Understanding on nuclear safety that created a framework for cooperation in areas such as emergency preparedness. Yet, of all the different venues in which the EU operates to secure its energy interests, one stands out: the Energy Community.

The Energy Community

The European Council identifies the Energy Community as one of the EU's most successful foreign policy frameworks (Council 2013). Its entry into force in 2006 was the culmination of years of EU effort to strengthen the rule of law on energy issues among the EU's main energy-supplying partners, particularly as pertains to trade and investments. It is an agreement between the EU and its south-eastern neighbours, several of which were slated to become member states (2006/500/EC). The idea behind it began in 1991 when Dutch Prime Minister Ruud Lubbers proposed establishing a European Energy Community. It took its first form with the 1994 signing of the Energy Charter Treaty (ECT) and the Protocol on Energy Efficiency and Related Environmental Aspects (PEEREA). The EU approved the treaty and protocol in 1997 (Council 1998) and both entered into force in 1998. Since then, the treaty was amended once (2010) to include some trade aspects and, in 2012, its secretariat approved a plan for consolidation, expansion, and outreach (CONEXO) with the objectives of, inter alia, expanding its geographic reach by convincing states with observer status (such as China) to accede to the treaty. As of early 2014, the ECT had 54 signatories, including the EU and Euratom. On the basis of the working relationships established through the ECT, the EU agreed with nine parties (Albania, Bulgaria, Bosnia and Herzegovina, Croatia, Macedonia, Montenegro, Romania, Serbia, and Kosovo) in 2005 to create the Energy Community. Romania, Bulgaria, and Croatia have since joined the EU, while Moldova and Ukraine became full members in 2010 and 2011, respectively.

The Energy Community is a diplomatic success story for the EU, a model example of both what the EU wants to achieve in its external policy as well as how it goes about pursuing its interests. The purpose of the community is twofold. First, it aims to 'organize the relations between the Parties and create a legal and economic framework in relation to Network Energy', which it defines as 'the electricity and gas sectors falling within the scope of the European Community Directives 2003/54/EC and 2003/55/EC1' (Council 2006c: 19; see 2006b for the related Council decision). Second, it should guarantee stable markets and a common regulatory framework to deliver continuous energy supplies to and between its participants. In essence, the Energy Community extends the EU's *acquis communautaire* on energy, environment, competition, and renewables to its south-eastern neighbours, a perfect example of rule export. It establishes the principle of mutual assistance in times of stress and through key governing institutions, coordinates policies between them. It is in many regards a legal extension of the EU, and a necessary prerequisite for further enlargement.

Energy Community decisions are made by its Ministerial Council and Permanent High Level Group (PHLG), which prepares materials for the ministers, and each consists of two EU representatives. It has a secretariat, which reviews the implementation of community decisions, and an Energy Community Regulatory Board (ECRB) consisting of regulators from the contracting parties and officials from the European Commission who coordinate an exchange of information (and in some cases establishes regulations) on regional gas and electricity markets.

The Energy Community has tallied several achievements since coming into existence. Beyond its geographical extension to include countries outside the Balkan region (Moldova and Ukraine), national legislation of its participants are now largely in line with EU law. The treaty was also strengthened to include EU legislation on security of supply, energy efficiency, energy labelling, and renewable energy, and its participants began implementing the EU's third package on the internal energy market and the EU's 2009 directive to maintain emergency oil stocks. Most important of all, however, was the October 2013 decision by the Ministerial Council to extend the duration of the treaty for a further ten years (Energy Community 2013).

Additional bilateral and multilateral efforts

The EU's diplomatic activities in relation to energy also include efforts to form coordination groups among countries along its periphery. These are often, but not exclusively, made through the auspices of the EU's Neighbourhood Policy, which it launched in 2003 in order to smooth relations with its neighbours in light of the 2004 enlargement. Among the most relevant to energy are the Euro-Mediterranean Energy Partnership and the various energy aspects of its partnerships and cooperation agreements (PCAs). As of March 2014, the EU maintained PCAs with Russia, the countries of Eastern Europe, the Southern Caucasus, and Central Asia (Council/Commission 1997, 1998a–b, 1999, 1999a–e, 2009). PCAs cover a broad range of political and economic reforms that aid in the transition to a market economy. To varying degrees they also contain energy chapters.

Through the Euro-Mediterranean Energy Partnership, (EUROMED; now known as the Union for the Mediterranean, or UFM), the EU promoted the Mediterranean Solar Plan (20 GW of regional solar power generation capacities by 2020) providing access to over €5 million through its Neighbourhood Investment Facility (European Commission 2009f), adding a technical assistance programme in 2010, and approving an additional €5 million in 2012 (European Commission 2012g). The EU's ongoing efforts to support 'the gradual establishment of a Mediterranean electricity regional market' constitute an essential component of its efforts to diversify its energy supply (European Commission 2014c). The EU likewise supports projects between the EU and the Littoral states of the Black and Caspian Seas, particularly via the Interstate Oil and Gas Transport to Europe platform (INOGATE), which it funds via the European Neighbourhood and Partnership Instrument (ENPI). Working within the frameworks of the Baku Initiative (which was finalized in 2006 and seeks, among other things, to harmonize regional energy markets and spur investments) and the Eastern Partnership (a venue for trade discussions and travel agreements), INOGATE provides a framework for cooperation in the areas of oil and gas, electricity, renewable energy, and energy efficiency between regional partners. Since its inception in 1996, INOGATE has completed some 70 projects, including ones closely related to EU energy security, such as promoting best practices and technologies used to detect, measure, and reduce gas losses from pipelines (European Commission 2006d). On the oil front, the EU helped set up the Joint Oil Data Initiative (JODI), a database of oil market players and transactions, as a method to bring transparency to one of the least transparent markets in the world. JODI is now a permanent reporting mechanism with 90 member economies, representing 90% of global oil supply and demand, that is, a platform that substantially enhances stability in the international oil market, particularly for investors.

As the EU shares a number of common principles with the US (open, transparent, and competitive global energy markets), and because it is not dependent on the US for oil or gas, its energy cooperation with Washington is a great deal simpler than it is with its immediate neighbours. In 2009, Washington proposed and Brussels agreed to establish the EU–US Energy Council to coordinate efforts on energy security, policy, and research and development in new technologies, including the sustainable development of biofuels and biomass, new materials, and Smart Grids (EEAS 2009, 2009a). They also agreed to cooperate on building up the Euro-Mediterranean Gas and Electricity Ring, as well as help create a new Southern energy corridor to the EU. Its meetings include the US Secretaries of State and Energy and the EU Commissioners for External Relations, Energy, and Science and Research, as well as the acting Commission President and High Representative. Although the number of specific energy-related agreements are so far few (e.g. energy appliance labelling), energy cooperation between the EU and the US led to the early development of discussion on raw materials and energy as part of their bilateral TTIP talks. The true measure of EU–US energy relations, however, stands to change in the coming decades as the US continues to press forward with the development and export of liquefied natural gas, which, if it comes to fruition, could substantially enhance the EU's long-term energy prospects.

The EU's most important bilateral diplomatic efforts focus on Russia through the EU–Russia Energy Dialogue. Its success has been limited. Launched in 2000, the Dialogue focuses on improving the investment climate in the Russian energy sector, and to secure and expand infrastructure in order to ensure continued production and delivery of sufficient supplies to the European energy market. Despite the relatively high-level communication and quantity of interactions, however, the process resembles what one expert called 'travelling without moving' (Talseh 2012).

As part of the Dialogue, the EU and Russia established bilateral fora (e.g. the EU–Russia Partnership for Modernisation (2010), Council (of the European Union) (2010)), a shared agenda for economic reform, and building a sustainable low-carbon economy. The EU and Russia also cooperate in the Gas Advisory Council and regularly exchange vital energy-related data through an early warning mechanism (see below). The Gas Advisory Council, which held its first meeting in November 2011, brings EU and Russian gas companies and experts together to assess and make recommendations for the long-term cooperation in the area of gas supplies. At their 8th meeting in November 2013, the two parties concluded, inter alia, that in order for the EU to 'meet its 20-20-20 targets', Russian gas supplies 'may not decline in any scenario' (GAC 2013). Meanwhile, following the 2009 gas disruptions resulting from a pricing dispute between Moscow and Kiev, the EU and Russia established an early warning mechanism to identify risks before they occur and ensure rapid responses in the event of emergencies (European Commission 2009g). The degree to which the arrangement works, however, remains untested. In February 2011, the two updated their agreement to provide for joint actions to be taken in the event of an emergency, including setting up special monitoring groups and notification procedures. Like its predecessor, however, the agreement is not binding and 'does not constitute an international agreement or other legally binding document', an emblematic and recurrent problem in the energy relationship between the EU and its most important supplier (European Commission 2011l).

Pipelines and their politics

Pipelines delivering natural gas and oil to the EU constitute its most important economic lifeline. The politics and economics of routes, distances, and capacities are some of the most complex and important features of an energy policy's external dimension. Unlike shipped commodities, pipelines create long-term, semi-permanent relationships between suppliers and consumers that lock them, de facto, into a marriage of convenience. Indeed, as one energy observer astutely noted, 'control of pipeline routes is almost as important as the control over the resources that flow through them' (Karbuz 2010). Once a pipeline route is up and running, it quickly becomes a lifeline that is difficult and risky to cut, and there are always winners and losers in the game. For example, by building the Nord Stream pipeline, Germany secured a reliable supply of up to 55 bcm (eventually) of Russian gas per annum for half a century, but in the process reduced the relevance of Ukraine, bypassed Poland, and added considerable constraints on German foreign policy, and thus that of the EU, vis-à-vis its relations with Moscow. The most important factor in determining a pipeline's economic and political value is its maximum capacity, because upgrading reception terminals or adding storage facilities is far less expensive and complicated than laying new lines.

Used to maximum capacity every day of the year, the EU's existing infrastructure of natural gas import pipelines could deliver approximately 13,400 GWh per day (or circa 405 bcm of natural gas every year) (ENTSOG 2013; see Table 7.1). Existing LNG terminals can add another 192 bcm of natural gas to that calculation, for a grand total of circa 597 bcm per annum (see Table 7.2), more than enough to meet all of the EU's 2012 annual gas consumption needs (circa 435 bcm). These maximum capacities obviously are never fully realized. Gas demand rises and falls with the weather and the competitive price of alternatives, such as cheap coal. Pipelines regularly close down for maintenance. In its Ten-Year Network Development Plan (TYNDP), ENTSOG notes that it evaluates pipelines and related projects on the basis of the adaptability and resilience of the European gas system, the dependency of the EU's gas zones on specific supply sources, and the capability of the network to enable supply diversification (ENTSOG 2013a).

At issue, then, is not so much import capacity, but rather the distribution of delivery capacities, which creates political and economic

Table 7.1 *EU gas import infrastructure (2013)*

Country of Origin [via transit state] (popular names)	*Technical, physical capacity (GWh/d) (ENTSOG)*	*Capacity (bcm/y)*
From NORWAY (Vesterled, Flags NLGP, Fuka & Sage, Europipe I & I, Norpipe System, Zeepipe, Franpipe, and Langeled South)	4,234.4	128.44
From NORTH AFRICA (Maghreb–Europe, Medgaz, Transmed, and Green Stream)	2,155.0	65.36
From RUSSIA [Direct] (Multipile Offshoots of Northern Lights and Yamal–Europe I, Nord Stream)	1,601.1	48.56
From RUSSIA [via BELARUS] (Yamal–Europe I and Northern Lights)	1,410.8	39.53
From RUSSIA [via UKRAINE]	4,022.2	122.00
From TURKEY	60.5	1.83
Grand Total	13,484.0	405.72

Notes: Conversion from GWh/d to bcm = GWh/d * 0.0831/1,000*365 based on IGU (2012) *Natural Gas Conversion Pocketbook*.

Figures exclude gas capacity for incoming lines designed solely to transit gas elsewhere, such as Turkey.

Source: ENTSOG (2013)

risk. Simply stated, EU pipeline politics is more about who delivers how much at which price, rather than the urgent need to get more.

Europe's pipeline infrastructure is characterized by the geographic location of the reserves to which it connects. These include the North Sea gas and oil fields, various Russian fields, Caspian and Central Asian fields, and the reserves of North Africa and the Middle East. The resources in these regions are accessed via three large corridors. One that runs southward from the Norwegian and North Seas to the UK, Scandinavia, and the continent. A second

Table 7.2 *EU LNG import infrastructure (2013)*

	Import terminals				Import capacity (bm^3 (N)/year)				Actual imports (bcm, 2012)***	2012 EU gas imports (%)°
	Exist.	Const.	Plan*	Tot.	Exist.	Const.	Plan**	Total		
March 2014 European LNG Import Capacity										
Belgium	1			1	9.0			9.0	4.28	1.2%
France	3	1	1	5	23.8	13.0	8.0	44.8	8.73	2.5%
Greece	1		2	3	5.3		5.6	10.9	1.45	0.4%
Italy	2	1	7	10	11.0	3.8	53.0	67.7	7.09	2.0%
Netherlands	1				12.0			12.0	2.18	0.6%
Portugal	1			1	7.9			7.9	2.33	0.7%
Spain	6	3		9	60.1	9.6		69.7	21.12	6.0%
Sweden	1		2	3	0.5		0.8	1.3		0.0%
Turkey	2		2	4	12.2		11.3	23.5	7.8	2.2%

(*Continued*)

Table 7.2 *EU LNG import infrastructure (2013) (Continued)*

	Import terminals				Import capacity (bm^3 (N)/year)				Actual imports (bcm, 2012)***	2012 EU gas imports (%)°
	Exist.	Const.	Plan*	Tot.	Exist.	Const.	Plan**	Total		
UK	4		2	6	51.1		18.0	69.1	14.23	4.1%
	22	**5**	**16**	**42**	**192.81**	**26.35**	**96.7**	**315.86**	**69.21**	**19.8%**
Plus planned newcomers: Future European LNG Import Capacity										
Albania			1	1			8	8		
Croatia			2	2			12	12		
Cyprus			1	1			0.7	0.7		
Denmark			1	1			::	::		
Estonia			2	2			5.5	5.5		
Finland			1	1			2.9	2.9		
Germany			1	1			10	10		
Ireland			1	1			3	3		

Latvia	1	1		5	5
Lithuania	1	1		4	4
Malta	1	1		2	2
Poland	1	1		5	5
Romania	1	1		2	2
Ukraine	1	1		5	5
Totals					
	32	**58**		**161.8**	**380.96**

Source: Belkin et al. (2013): 3; ECS (2008): 103; ENTSOG (2013); GLE (2015): LNG MAP; IGas (2013): 11, 13.

Notes: *LNG import terminals under study or proposed; **Planned sources (under consideration) taken from ECS (2008), Table 4, p. 103 (Germany and Cyprus) and from GLE (2015): LNG MAP; ***Data from IGU (2013): 11; ° In 2012, the EU imported circa 350 bcm in total according to Belkin (2013).

running northward from North Africa to Spain and Italy. A third running eastward from, or through, Russia and the former USSR, and is best divided into two sub-corridors. The northern sub-corridor carries oil and gas from Russia directly to Germany and the Baltic States, or transits through either Belarus or Ukraine. The southern one brings those resources from the Caspian, Black Sea, Lower Central Asia, and the north-western Middle East to south-east Europe, via Turkey or Russia. Efforts are underway to build a fourth corridor connecting to the same south-eastern regions, but bypassing Russia. The EU repeatedly identifies such efforts as priority projects and, as of early 2014, there were two projects: TANAP (the non-Russian plan) that was approved and is moving towards construction; and South Stream (the Russian-controlled alternative) that was indefinitely suspended in December 2014 due to increasing tension between Russia and the EU.

The pipeline strategy and its controversies

By accessing enough suppliers as directly as possible, the EU is hoping that increased supplier competition will keep supplies flowing and prices low. The problem begins with the prioritization between direct access and diversity. The fact is that the EU needs both to enhance its security of supply. Yet that does not stop debate. For example, the Nabucco project, originally a favourite in Brussels and designed to bring Caspian, Central Eurasian, and Middle Eastern gas to Central Europe without Russian intervention, ultimately had to be truncated (now it is Nabucco West), as neither suppliers nor routes could be secured in a timely and cost-effective manner. Meanwhile, Russia implemented plans to increase Russian supplies and bypass Ukraine through bilateral arrangements between Moscow and individual EU member states. It successfully built the Nord Stream to Germany and, in early 2014, was promoting its proposed South Stream project to secure a role in the Southern energy corridor. Despite the rhetoric, the choice of lines is generally not a zero-sum game. For example, South Stream was eventually superseded by Turkish Stream. Every project offers advantages and disadvantages, and if every planned pipeline were built, Europe would significantly enhance both its import capacity as well as reliability of those supplies.

Most existing or planned pipelines, such as Turkish Stream and Nord Stream, are upstream-driven, where specific supplier companies or countries such as Russia propose, finance, and build pipelines to provide outlets for their own resources. Russia's plans, first for South Stream and subsequently Turkish Stream, complicate Europe's plans to diversify away from Russian resources. Part of that complication is the fact that there are a lot of Western companies driving and/or participating in the projects, which in turn informs us that such projects are inevitably as much about money as they are about politics. Still, given Russia's pursuit of available resources and its active plans to bypass transit states to deliver gas to Europe, it is understandable that many saw the South Stream project as a Russian power-grab.

The downside of South Stream was that it was not at all clear that the pipeline actually would have brought any new supplies to the EU, but rather just serve as a detour for existing Russian supplies currently flowing through Ukraine. Indeed, South Stream would have strengthened, not diminished, Russia's strategic position vis-à-vis the EU. Nevertheless, the addition of new routes directly from Russia into the EU offers distinct advantages over the current network configuration, the most important being the reduction of transit bottlenecks and the direct supply of gas to Europe. According to Russia's then Prime Minister Vladimir Putin, Russia has enough gas to supply Europe 'for practically a century' (Stratfor 2009). The failure of the South Stream project was a major political defeat for Moscow and a regulatory victory for the EU. The message was clear: you want to play in our backyard, then play by our rules. However, Russia's subsequent project, Turkish Stream, opens up a whole new can of worms for Europe. For example, if Russian gas previously transiting Ukraine ends up in a Turkish hub to be resold, then Europe could end up paying more for the same gas. Likewise, any new Turkish–Russian gas deal brings gas to a location where Europe cannot properly access it because the necessary infrastructure is not yet in place.

Russia made a powerful statement about its long-term reliability and plans by building the Nord Stream, a joint Russian–German project to deliver up to 55 bcm of Russian gas per year directly to Germany via two pipelines spanning the Baltic Sea. The construction costs of the line (from Vyborg to Greifswald) outstripped estimates for upgrading pipelines running through Belarus and Poland, revealing that both Russia and Germany think that Belarus and Ukraine negatively affect the reliability of Russian supplies. Nord Stream mitigated some of that problem, but the bulk of Russian gas supplies continue to flow through Ukraine. The project also inspired strong opposition in Europe. For example, Poland's Defence Minister, Radek Sikorski, compared the Russian–German plan to deliberately route around Poland and the Baltic countries to the pact between Hitler and Stalin (Spiegel 2009).

Europe's energy-corridor map is changing. From the north, new sources from the Arctic and from the south new sources from the Caspian and Middle East will make their way to Europe. Much of it will arrive via Russia, some directly and some indirectly. Yet, as the Nord Stream project demonstrated, more gas and new routes are always in the Union's interest. Indeed, now built, it is hard to argue that Nord Stream did not enhance EU energy security.

The North-Western Corridor from Norway

The Northern route supplies circa one-third of the EU's current gas imports and includes the gas and oil fields of the North Sea, Norwegian Sea, and, in the near future, the Arctic Sea. Norway is the most important Northern route operator. North Sea gas and oil deposits were known to exist in the late nineteenth century (Pratt 2007), but large-scale production only began in the 1960s, when the North Sea coastal states (UK, Sweden, Germany, Belgium, Denmark, and the Netherlands) agreed to the seabed's dividing lines. The Norwegian sector contains the largest reserves.

Norwegian energy exports are essential to EU energy-supply security. They flow through a pipeline network operated by the country's state-owned monopoly Gassco, which took over as the operator of Norway's natural gas transportation system in 2001 from Statoil, the national energy company. According to BP (2013, 2014), Norway provided 28% of EU gas imports in 2012 (up from 24% in 2006) and 12% of its oil imports (down from 16% in 2006). Most importantly, Norway's deliveries to the EU have been consistent for decades, making it Europe's most reliable supplier. It operates 22 different pipelines delivering gas and oil via thousands of kilometres of pipeline to the UK, Germany, Belgium, France, and Denmark (Gassco 2014). It delivered around 106 bcm of natural gas to the EU in 2012 (BP 2013, 2014), but its complex undersea pipeline network could move as much as 120 bcm per year. Moreover, its production should remain steady in the coming years, as new discoveries and investments are extending production in existing fields.

Gassco's complex pipeline network connects various offshore fields to riser platforms in the North Sea; the company provides an interactive map on its website (Gassco 2014). Two major lines (Langeled and Vesterled) deliver gas to the UK, three (Norpipe, Europipe, and Europipe II) deliver gas to Germany, and two (Zeepipe and Franpipe) deliver to Belgium and France, respectively. Norway suspended plans to build a pipeline (Skanled) to bring circa 20 bcm/y to Denmark, Sweden, and Finland in 2009 due to concerns over long-term demand, a by-product of Europe's economic recession.

North Sea production plateaued in the 1990s and has declined annually since 2001. In response to North Sea declines, Norway has been exploring options further north in the Norwegian, southern Barents, and Arctic Seas. According to the Norwegian Petroleum Directorate (2013: 28), the Norwegian Sea may contain as much as 680 bcm of natural gas (plus an additional 445 bcm yet undiscovered) and 147 mcm of oil (plus another 300 mcm yet undiscovered). Unlike the exhaustively developed North Sea, Norwegian Sea resources are still under development. The first field came online in 1993 and a conglomerate led by BP started pumping from the Skarv field in December 2012, which is believed to hold 100 million barrels of oil and over 48 bcm of rich gas. Development of the Arctic's oil and gas resources is a key component of Norway's strategy and fits well into Europe's strategic energy vision. Norway believes that there may be as much as 550 bcm of undiscovered natural gas reserves and 2.6 billion barrels of undiscovered oil in their part of the Arctic region (Norwegian Petroleum Directorate 2008, 2012, 2013, 2014). As in the case of the Norwegian Sea resources, tankers, not pipelines, will deliver almost all resources extracted from beneath the Arctic's icy waters. This is an important point, since Europe's tanker and refinery network is well established, so no special actions will have to be taken to incorporate any new oil deliveries. In order to profit from any new gas production, however, the EU will have to significantly expand its LNG import facilities.

Norway is not alone in the development of Barents Sea resources. Its main competitor, Russia, controls the largest share of resources in the region of the South Barents Sea, as well as the Siberian Basin underneath the Kara Sea. While it is conceivable that Russia might consider future pipelines from the region to its north-western pipeline network, and thus on to Europe, the more likely scenario is their establishment of large-scale LNG-processing facilities dotted across the region.

The South-Western Corridor from North Africa

The EU's South-Western energy corridor is small, simple, and transparent. Europe's pipeline interests in the south are exclusively dedicated to natural gas. Algeria is the EU's third largest natural gas supplier delivering circa 50 bcm in 2012, or 11% of EU gas imports (BP 2013). Most Algerian gas is exported via pipeline (70%) of which two-thirds go to Italy, another third flows to Spain. The remaining 30% travels almost exclusively to the EU and Turkey, and does so in the form of LNG. North Africa's other gas suppliers, Libya and Egypt, delivered less in recent years, as both underwent political turmoil. Nonetheless, Libya was able to deliver 6.5 bcm to Italy in 2012. The relatively small share of North African gas in EU consumption is a positive sign for the short to medium term, because the region's gas (8 tcm of gas, 4.3% of the world's proven reserves) remains largely untapped (BP 2013). Given its proximity, North Africa could become a reliable long-term source of natural gas. Some believe that Algeria alone could eventually export as much as 130 bcm/y by 2020 (Poynter 2007), although such figures seem overly optimistic given past and current trends.

There are four operating pipeline links from North Africa to the EU with a maximum annual capacity to deliver just over 57 bcm. They all traverse the Mediterranean Sea and include: the oldest and largest, Transmed, now called the Enrico Mattei Gasline (Algeria to Italy via Tunisia); the Maghreb–Europe (MEG; Algeria to Spain via Morocco); Green Stream (Libya to Sicily and onwards to the Italian mainland); and the Medgaz (Algeria to Spain). However, Green Stream was closed for several months during 2011, and for a month in 2013, due to unrest in Libya. The European Union identified another pipeline, the Galsi, to run from Algeria to Italy over Sardinia, as a priority project in its Trans-European Energy Network, and even allocated €120 million to finance its construction. However, the project was stalled in early 2014 as 'weak demand' combined with 'environmental concerns, economic issues, and anticipated supplies from the southern corridor projects' undermined its viability (Smith 2014).

Completion of the Arab Gas Pipeline, linking together Egypt, Jordan, Syria, Turkey, and Iraq, eventually may allow some additional North African gas to find its way into Europe via the Southern Corridor. Given the political turmoil in the area, however, prospects for such developments seem slim anytime soon. Indeed, although Libya's natural gas exports picked up again in 2012, its only LNG export facility remained offline at the time of this writing (2014), and most Egyptian natural gas remained bottled up at home for the country's thriving domestic market, where gas is subsidized and widely used for cooking, heating, and public transportation. Nevertheless, Egypt is making substantial investments in LNG export facilities and, while neither Egypt nor Libya will be able to match Algerian production, together they could add significant quantities to the region's exports in the not-too-distant future, easing pressure on Algeria and providing the EU with additional options.

The North-Eastern Corridor from Russia and the FSU (oil)

Russia is the EU's main supplier of oil and gas coming from the east, delivering both via a complex network of pipelines that, with the notable exceptions of the new Nord Stream and currently planned Turkish Stream lines, is ageing and in need of both repairs and upgrades, worsening the risks of EU dependence. Also, the network transits Ukraine and Belarus, leaving Russia and the EU at the mercy of disruptions and contract disputes. Although upgrades to the lines might resolve some of the maintenance issues, investments in repairing the network will not resolve the instability derived from the transit issue. Russia's dependence on transit countries plays an important role in driving Russian desires to establish independently managed, direct links to the EU gas and oil markets.

Many of the gas lines converge in Ukraine, and then travel westward. Most of the older Soviet lines originate in the Urengoy, Yamburg, and Medvezhye super-giant gas fields, the more recently developed Orenburg field, or along the Yamal peninsula. The pipeline that runs from Orenburg to Uzhgorod is also referred to by some as the Brotherhood-Union line, which is confusing, because the Soyuz (which in Russian broadly translates into unity) starts in Orenburg, and the Brotherhood or Friendship line (Druzba) starts in Urengoy, far to the north. The Soyuz line follows an entirely different route, although they do both come together as export links to Europe via Ukraine. The Druzba dates back to Soviet times. It was constructed to be a domestic line that carried Soviet–Ukrainian gas from the 'Carpathian fields south of Lvov' to Czechoslovakia (Kamm 1968).

An average of about 482,000 tonnes of oil a day flowed westward into the EU in 2011, based on annual figures (Eurostat 2013). Russia's oil is exported via pipelines, exclusively operated by the Russian state-controlled oil transport company Transneft, converging in Belarus and branching off into north and south branches. The north branch runs from Belarus into Poland and onto Germany, which received, respectively, circa 97 and 60 thousand tonnes a day in 2011. The south branch runs through Ukraine and splits further into two lines, one running through Slovakia to the Czech Republic, and the other through Hungary, connecting to the Adria pipeline, which brings oil to EU neighbours Serbia and Croatia. Another offshoot connects Latvia and Lithuania. An accident along that line led to a stoppage in 2006; one that intriguingly coincided with the sale of refineries in both countries to non-Russian parties.

Three additional oil pipelines, the Baltic Pipeline System (BPS), the Odessa–Brody (OB), and the Caspian Pipeline Consortium (CPC), also bear mentioning. The BPS delivers over 1.4 million barrels of oil per day, via two lines from the western Siberian, Ural, and Timan–Pechora regions to Primorsk in the Russian Gulf of Finland. Tankers then take supplies from Primorsk to markets worldwide, the closest being the Nordic European states. The domestic route frees Russia from transit costs across the Baltic States, lowering oil transportation costs by as much as a billion dollars a year (Ibrahimov 2007), and is part of Russia's strategy to free itself from problems associated with transit states that have hurt the country's reputation as a reliable supplier. Odessa–Brody is a Ukrainian domestic line designed to carry oil from the Ukrainian port city of Odessa to Brody near the border of Poland. Built in 2002, the original idea was to bring Caspian oil from Kazakhstan to the Polish Baltic Sea port of Gdansk. However, a lack of supply and Ukraine's failure to find shippers left the pipeline empty for two years following its completion. Odessa–Brody now ships Russian oil south, instead of Kazakh oil north, demonstrating how a failure to feed a line literally can reverse the winner–loser calculation. The CPC runs from western Kazakhstan to the Russian Black Sea port of Novorossiysk, and from there, oil is shipped to Europe and other markets. Until the opening of the CPC, Druzba was Kazakhstan's main oil export route to Europe.

A few smaller, but nonetheless important oil pipelines, traverse or bypass Russia altogether, the most famous being the Baku–Tbilisi–Ceyhan (BTC) Crude Oil Pipeline. The BTC can carry as much as a million barrels a day of Azeri oil to Ceyhan, Turkey, traversing Georgia. The BTC is, however, an anomaly. Most oil pipelines from the Caspian region still flow to Russia, including the Northern Early (Northern Route Export Pipeline), flowing from

Baku to Novorossiysk, and the Western Early (Western Route Export Pipeline), flowing from Azerbaijan to Georgia's Black Sea port of Supsa, which together carry another 800 thousand barrels a day. Some projects, such as Burgas–Alexandroupolis Crude Oil Pipeline (BAP) – a Russian, Greek, and Bulgarian initiative designed to bypass Turkey's Bosporus and Dardanelle Straits – failed to take off, while others, such as the 900-barrel-a-day Albanian–Macedonian–Bulgarian Oil Corporation (AMBO) line and the Pan-European Oil Pipeline (PEOP) seem to be moving forward despite occasional political obstacles.

Ceyhan, Turkey currently serves as the terminal for both the BTC Pipeline and another pipeline from Iraq called the Kirkuk–Ceyhan, or Iraq–Turkey Pipeline (ITP). The ITP served as the main conduit of Iraqi oil throughout the Iran–Iraq war, when shipments by tanker out of the Persian Gulf were in danger, and is the main export link for Iraq's northern oil fields. The ITP plays an important role in Turkey's plans to become a link between the Middle East and Europe, and strongly figures in Turkey's plans to make Ceyhan an international energy-export hub. In the future, a parallel line could carry northern Iraqi gas to Ceyhan, further increasing the importance of the port and strengthening Turkey's hands in its relations with the EU.

The North-Eastern Corridor from Russia and the FSU (natural gas)

Russian natural gas is pumped into its vast gas pipeline network called the Unified Gas Supply System of Russia (UGSS). The most important links of concern to Europe are the Soviet-era Brotherhood and the recently constructed Yamal–Europe pipelines. These two pipelines, together with a network of over a hundred smaller lines, meet in either western Ukraine or Belarus where they then connect onwards to Europe. Several Russian domestic links are essential to the country's export capacity. These include the Northern Lights line (western Siberia to Ukraine); the Soyuz (Urengoy to Uzhgorod), after heading south to Orenburg to pick up gas from fields near the Kazhak border; and the Progress, which is another westward-flowing interconnection that ties into the Soyuz–Brotherhood feed. Russia is planning to extend the Yamal–Europe Pipeline to run from the Yamal peninsula in Russia's Arctic north to Frankfurt (Oder) near the Polish–German border, but as of 2013, it was still in the feasibility study stage.

In 2013, 4.8 million terajoules (circa 126 bcm) of Russian gas flowed westward into the EU (Eurostat 2015d). Still, Russia's dependence on transit countries is a latent problem for the Northern East–West gas corridor. Of the primary gas arteries, the Brotherhood–Europe interconnection, the first line to deliver Soviet gas to Western Europe, following its extension from Czechoslovakia in 1967, to Austria in 1968, is the oldest (Gazprom 2014: 51). From Ukraine, Russian gas flows into Slovakia and Hungary. From Slovakia, the route continues into the Czech Republic and on to Germany and France. Additional branches shoot off to southern and south-eastern Europe, one route taking gas through Austria to southern Germany and Italy. Other Brotherhood branches supply Romania, Bulgaria, Greece, Serbia, Bosnia, Macedonia, and Turkey. The most important part of the Soyuz export pipeline, commissioned in the late 1970s, runs from Orenburg in the southern Ural region to Ukraine, where it hooks up with the Brotherhood lines to Europe. The Yamal–Europe Pipeline, also known as Yamal-I, has been operational since the late 1990s and brings Russian gas from the Urengoy fields in western Siberia, across Belarus, into Poland, and on to Germany. The interconnection is located in Torzhok, north-west of Moscow, and together, the north-central Russian links are known as the Urengoy–Nadym–Peregrebnoye–Ukhta–Torzhok transmission system. It is entirely domestic, and vital to reliable Russian exports to Europe.

Although its network is complex, there are only four routes over which Russia currently provides gas to Europe or Turkey. The first, and largest, goes through Ukraine. The second traverses Belarus. The third crosses the Black Sea to Turkey, and the fourth, and newest, route travels under the Baltic Sea directly to Germany (Nord Stream). Transit is the main bottleneck in Russia's gas export network to the EU. Some 80% of Russian gas, bound for Europe, travels through one country, Ukraine, which leaves little margin for error. When problems do occur between Russia and either Ukraine or Belarus, irrespective of the cause, Russia looks like an unreliable supplier. While EU members such as Germany and Poland could theoretically receive gas through the Yamal–Europe Pipeline that crosses Belarus, the latter has a limited throughput capacity of about 30 bcm compared to the more than 140 which could flow westward through Ukraine. Pricing disputes, maintenance problems, and the interdependence between the EU and Russia on transit states translates into instability in the EU–Russia energy relationship.

Recognizing this fundamental problem, the EU is actively pursuing alternative suppliers, while Russia is scrambling to establish alternative routes. Until either succeeds, neither will be content.

While Europe and Russia struggle to balance their specific needs, transit states Georgia and Ukraine have been pushing for a new pipeline to be built that would allow both countries to become direct transit states for Caspian gas, bypassing Russia altogether. Known as the White Stream Gas Pipeline, the project would carry at first 8 bcm, and ultimately 32 bcm, of natural gas northward from Azerbaijan and Turkmenistan via a submarine gas pipeline, across the Black Sea directly to Romania, or first through Ukraine. If the route flows through Ukraine, the country would be able to reduce its reliance on Russia as its only source of Turkmen gas supplies. If, instead, the undersea line flows all the way from Georgia's Black Sea coast to Romania, the EU would have only one transit state (Georgia) between itself and vital Caspian gas resources. The White Stream project illustrates the importance of transit states in Europe's East–West corridor, but it also highlights the importance of the southern plane of that corridor.

The South-Eastern Corridor (old and new)

The difference between the transit problems of the north and those of the south are not only the sources tapped but also the countries traversed. In the north, gas from Russia's Arctic, west Siberian, and Ural regions all transit Ukraine and Belarus before coming to Europe. In the south, gas flowing westward to Europe from the Caspian and Central Asia traverse either Russia or Turkey. Central Asia's largest gas network is the Russian-controlled Central Asia-Center (CAC) Pipeline. Turkmenistan, Kazakhstan, and Uzbekistan, all of which are rich in energy resources, send most of their gas through the CAC's five-line network, which constitute an important gas supply route for the EU. The eastern branch (CAC-1, 2, 4, and 5) starts in Turkmenistan, picks up more gas in Uzbekistan, and then heads north-west, via Kazakhstan, to Alexandrov Gay on the Russian side of the Kazakh border, where it connects to the Soyuz and Orenburg–Novopskov pipelines that run north-west to Moscow. The CAC's western branch (CAC-3) begins near Turkmenbashi, runs north to Ekarem, and then along the Turkmenistan–Iranian border to Uzen in Kazakhstan, where it connects to the eastern branch (IEA 1998). In 2007, Russia agreed with

Turkmenistan and Kazakhstan to build another 20-bcm pipeline parallel to the northern part of the CAC-3 line, but progress has been slow and erratic, and for some at the time (MEES 2008), was viewed as a move by Russia to secure natural gas supplies from Turkmenistan, thus preventing alternative trans-Caspian solutions that exclude Russia. Because of the Caspian region's geographic position and history, Russia's role as a transit state is similar to the role played by Ukraine and Belarus, with two important distinctions: Russia is both a massive consumer and the region's undisputed suzerain. Thus, bypassing it might make economic sense for Central Asian suppliers, but it could also worsen relations with Moscow.

The Baku–Tbilisi–Erzurum (BTE) Pipeline bypasses Russia altogether and can deliver up to 9 bcm annually to Turkey, via Georgia. The line has been plagued by troubles from its inception, though, including supply problems from Azerbaijan and a two-day suspension in 2008 following Russia's invasion of Georgia. Importantly, the same source of the BTE, the Shah Deniz gas field, is also expected to feed the proposed TANAP gas pipeline across Turkey and connect to, inter alia, the EU's Nabucco West. Turkey, meanwhile, is increasingly becoming an important transit player in Europe's quest for natural gas. Since 2002, Turkey has been receiving additional supplies directly from Russia via the Blue Stream pipeline, which crosses under the Black Sea and bypasses the older transit routes of Moldova, Bulgaria, and Romania. An extension of the Blue Stream line is currently under consideration.

Gas started flowing (11.5 bcm in 2011) from Turkey to Greece via the Interconnector Turkey–Greece (TGI) in 2007. It is a foundational component of the EU's plans to create a new Southern energy corridor. The plan for these networks is to deliver gas from Azerbaijan's Shah Deniz gas field across Turkey and Greece into Italy and Bulgaria. Additional supplies will have to come from other Caspian sources. Another pipeline, called the Trans-Adriatic Pipeline (TAP), will branch off the TGI in northern Greece and carry gas through Albania, on route to Italy, via a submarine line through the Adriatic Sea. In June 2013, the Shah Deniz Consortium selected TAP as their preferred transportation route to deliver Caspian gas to Europe.

The development of the Southern Corridor positions Turkey as a major player in Europe's future energy security. Turkey also offers an important, if not controversial, link to Iran. The East Anatolian

Natural Gas Pipeline (EANGP) has been moving gas from Tabriz in Iran to Erzurum since December 2001. Although the two countries regularly disagree over price, volume, and reliability (Iran halted shipments in order to meet domestic demand over three consecutive winters), Tehran sends as much as 8 bcm to Turkey annually (MEES 2008). Iran is vastly rich in gas reserves, but needs large investments to increase its exports, and first has to resolve its political disputes over its nuclear programme before it can become a regular supplier of gas to the EU. If and when it does, Turkey will play a central role.

The maturing LNG option

Pipelines are only part of the EU's security of supply network. Europe's natural gas is increasingly arriving by ship in the form of Liquefied Natural Gas (LNG). Liquefied Natural Gas (LNG) is the technical term for super-cooled natural gas transported by ship and delivered to storage and regasification terminals at distant ports. As of 2014, there were 22 functioning LNG terminals in the nine member states and Turkey that were capable of importing 192 billion cubic meters and storing 8.6 bcm on any given day (see Table 7.2). Taking into account five more terminals under construction, and plans to build another 32 in 13 additional member states plus Ukraine, those figures are expected to explode to 380 and 16.5 bcm by 2022 (Groenendijk 2014). Still, it was not fully clear in early 2014 whether an expected boom in US LNG exports (Belkin et al. 2013) could overcome domestic economic and political obstacles and, thus, add enough supply to the market to significantly lower prices (EIA 2012). In fact, US LNG exports in 2012 were a negligible 0.6 bcm (BP 2013, 2014), and EU imports actually declined to 45 bcm in 2013, almost half the figure (86.3) in 2010 (Groenendijk 2014). The decline in LNG imports was largely due to a decline in EU demand, resulting from an economic slowdown and higher import prices, as well as demand in Asia. Despite the decline, however, LNG still accounted for 10% of all EU gas consumed in 2012, and the EU is actively and materially supporting the growth of LNG in its energy market by approving and supporting new import terminals and backing a demonstration project to set up a chain of refuelling stations for heavy-duty, LNG-fuelled vehicles across four major European trucking routes (LNG Blue Corridors 2015).

LNG changes Europe's energy-dependence paradigm in two important ways. First, ship-delivered gas is free from overland transit politics, a factor that has so far played a largely negative role in supply pricing, diversification, and reliability. It also eliminates the need for eminent domain (forced seizure and compensation of land), adding competition to an otherwise narrow and oligopolistic market. Second, LNG deliveries open a new seaborne delivery corridor and expand the roster of available suppliers, thus enhancing diversity, while increasing supplies, and reliability. The most important advantage of LNG is its flexibility. Unlike pipelines, ships are more compliant to market needs, able to deliver gas from any origin to any coastal termination point. For exporters and shippers, LNG allows the matching of deliveries with market needs and profitability margins, which in turn enhances competition and establishes the foundation for active international spot and future markets. In short, LNG increases the options to diversify suppliers in ways that pipelines cannot. Also, because storage is an essential component of the process, LNG leads not only to increased diversity but also enhances emergency response capabilities.

The EU's LNG infrastructure can be divided between its northern and southern flanks. Five north-western member states import LNG (UK, Belgium, France, Sweden, and the Netherlands) and three more (Germany, Denmark, and Ireland) plan to do so, as do five north-eastern members (Estonia, Finland, Latvia, Lithuania, and Poland). Portugal, Spain, Greece, and Italy make up the four western importing members, and in the south-east another five member states (Albania, Croatia, Cyprus, Malta, and Romania) have plans in the works. Europe has been importing LNG since the mid-1960s, even before it began receiving externally supplied gas via pipeline. Algeria began delivering LNG to France and the United Kingdom in 1964, while the Netherlands only began exporting gas via pipeline in 1967. Since Spain and Portugal are located on the furthest end of the Norwegian and Russian pipeline networks, they began to invest heavily in LNG in the late 1990s, increasing Algerian imports and adding Nigerian LNG to their portfolios. In 2001, Spain surpassed France in imports and remains Europe's largest LNG importer, followed by the UK, France, Turkey, and Italy (IGU 2013).

Europe's LNG comes from a variety of sources. As distance from market affects price upwards, location of suppliers is important, meaning that the EU will tap resources generally found in the Atlantic and Middle Eastern markets. In 2012, the US, Trinidad & Tobago, Peru, Norway, Qatar, Algeria, Egypt, Equatorial Guinea, and Nigeria each delivered LNG to the European gas market. The largest supplier is Qatar. Since beginning to export the fuel in 1996, Qatar grew into the world's largest single exporter (77.4 million tonnes or 108 bcm) in 2012 (IGU 2013). The wealth of LNG in the Middle East provides opportunities for European energy security, but they are by no means a panacea. Despite its enormous resources, supplies from the region face serious obstacles, including growing domestic demand, subvention programs that hinder exports, and enormous buyer competition (Weems and Midani 2009). African suppliers are important partners too, but, unlike Qatar, whose exports continue to expand, theirs have remained stable. The global balance is likely to change further as new fields in the Arctic come online and US export terminals receive approval. Russia tried to develop its Arctic production options in the Barents Sea, but its Shtokman project, which it initiated in 1992, still does not produce LNG. Even if Arctic LNG does become a viable option in the near future, shipping will remain a problem, because developers will need new LNG carriers with ice-strengthened hulls and international cooperation to monitor sea ice.

Matched with increased liquefaction and an overcapacity to receive LNG, new flexibility is coming to the market. However, the capital-intensive nature of the industry, and the complications associated with gaining access to terminals, makes it difficult for new parties to join the fray; and since most LNG trade remains locked in fixed delivery contracts, most shippers are cautious to take their vessels out of permanent rotation, a development necessary for spot trading to work. Spot markets are being aided by the appearance of European gas hubs, adding further price competition and regionalization to the market. Belgium, France, the United Kingdom, and Austria each initiated their own hubs, creating greater opportunities and laying the foundation for a well-supplied future internal market. If such a market emerges, Europe will greatly enhance its energy security by seizing the upper hand in price-setting of a resource considered central to its plans for energy diversity and reliability. The LNG market is also becoming more transparent in Europe. Following the EU's increasing requirements, for example, GLE Europe voluntarily publishes weekly updates on daily LNG storage and imports at EU terminals (Gas LNG Europe 2015).

Finally, although domestic sales still account for most of the world's annual natural gas production, with pipelines accounting for most of the international trade, LNG is emerging as a vibrant and growing market. Global liquefaction capacity is increasing, as are the number of importers, terminals, and carriers. LNG plays a major role in Europe's attempts to achieve energy security and diversity. In the short term, the EU will continue to maintain its import overcapacity. That is a good thing if and when current sources of natural gas are interrupted; the EU can turn to LNG to make up for the shortfall.

Concluding remarks

EU external energy policy in the early twenty-first century is not what it was prior to the Lisbon Treaty. The Commission is increasingly taking the initiative, leading the member states under the flag of market-building. Although it certainly does not always work to the Commission's satisfaction, the EU increasingly looks and behaves like a more cohesive body in the area of its external energy relations, at least relative to what it was just a decade ago. Diplomacy is the primary tool of the external dimension, where interests are secured through multilateral frameworks, the most important being the Energy Community. EU efforts to expand the reach of the energy acquis through such regional institutions as well as bilateral dialogues and partnerships are helpful. However, without a healthy, diversified surplus of energy resources flowing into a functioning internal energy market, those multilateral efforts will continue to suffer under the strains of national interest.

The EU faces external energy problems rooted in the historical development of the member states' energy-import infrastructures and the long-term relationships they established with their suppliers. While this problem is not going to disappear anytime soon, the member states seem to agree that the single most important problem is the unbalanced dependence on Russian gas delivered by Russian-controlled pipelines. Import capacity is not the issue, but rather who delivers what to whom at which price; and that is what makes the external dimension so politically volatile. The EU is trying to both change Russian behaviour through partnerships and interaction as well as find new ways to get gas to Europe. It has been somewhat successful with the latter; not so much with the former. Diplomacy is helpful, but in an increasingly regionalizing world, it may be no match for market power and political ambition.

Chapter 8

Policy Challenges on the Horizon

As it tries to meet the energy demands of its consumers, the EU needs to resolve several policy challenges. Almost everyone in the Union agrees on the need for comprehensive energy security, but achieving anything close to it will require resolving a lot of unfinished business, both internally and externally. Issues in the internal dimension are comparatively easy to resolve once one identifies and bridges the gaps between national interests. The external dimension is more problematic, because the most severe issues, and those most likely to continue to divide the Union, are geopolitical in nature.

In keeping with our structured approach of the energy typology, we address these challenges by grouping them into the internal and external dimension, incorporating interdependent, multidimensional policies pertaining to environmental sustainability and climate change. Internally, the EU faces serious questions over such diverse matters as when, or even if, it will ever complete its internal energy market, which measures will be necessary to improve energy security, and when and how it will develop true solidarity to form a common energy policy for all the member states. Along the way, it must resolve open questions about the compatibility of creating a carbon-light economy and achieving comprehensive energy security. How economically viable are renewables after all? Are increasing climate change targets useful launching points for policy or rather, do they shroud out realizable alternatives that could have a stronger impact in the long run? Should the EU consider increasing indigenous fossil fuel exploration, in particular the highly controversial exploitation of shale gas? As all these questions illustrate, EU energy policy is plagued by the constant struggle between which direction to take and how to get there, as well as between the concurrent need for solidarity and the securing of national interests.

Even as it pursues resolutions to its internal problems, the EU and its member states must maintain existing import avenues and find new ones in order to both increase the quantity of supplies flowing into Europe as well as offset any possible disruptions. As this challenges the regional spheres of influence of other states, Russia in particular, it steps right into a morass of political trouble. Thus, we chart four external cases that reveal the geopolitical challenges through which EU policymakers must navigate: managing relations with Russia, gaining direct access to Caspian resources, defining its relationship with Turkey, and determining whether, and how, to jointly exploit the vast resources under the Arctic Sea.

The EU's internal challenges

Moving forward with the internal market

European institutions have generated enough documents emphasizing the merits of a well-integrated internal energy market to fill libraries. The idea and variations of its assumed value are almost commonplace in the statements of the European Commission, the Council, and the EP; and almost every stakeholder, from industry associations to consumer organizations, have long-since joined the chorus. The three liberalization packages brought tangible benefits to consumers by expanding competition and integrating previously physically detached energy markets. For example, the unbundling components freed consumers to choose their energy suppliers to an unprecedented degree throughout most of Europe, and some pilot projects, focusing on regional cooperation, established the so-called 'day-ahead market coupling', a mechanism that manages cross-border electricity flows from the Baltic States to the Iberian peninsula.

Despite all the progress, however, the speed of integration remains modest, seeming to accelerate only if external crises force the EU to act. It is not as though the challenges concerning the internal energy market are much different from other sectors of Europe's single market. As elsewhere, it is first and foremost a matter implementing existing laws. As the Commission has amply demonstrated, compliance rates are not indicative of a strong will among the member states to strengthen the single energy market. Indeed, while a harmonized legal framework (in Euro-speak, Network Codes) is being developed under the authority of ACER and network operators, progress is uneven between the electricity and the gas sector, as well as between regions. In addition to the uneven implementation of energy regulations, transparency is a major ingredient of a functional market. Without it, markets tend towards oligopolies, monopolies, and other distortions (e.g. so-called national champions). Important steps have been made to improve the transparency dimension, such as Regulation 1227/2011, which establishes rules on wholesale energy market integrity and transparency (EP/Council 2011). Another example concerns the establishment of a central information transparency platform for the publication of data in electricity markets in early 2015. Thus, it is not about the finalization of the single energy market as such – a running project ebbing and flowing along with changes in energy mixes, technological advancements, and societal mood – but rather the uniform application of existing rules, both in fact and in spirit.

The good news is that consumers reap the benefits of increased competition. In January 2014, the Commission reported that wholesale energy prices for electricity had dropped significantly, and that wholesale gas prices had remained stable since 2008, even though taxes and levies had risen. The bad news is that dismantling national champions stymies support for large-scale infrastructure projects. Establishing a joint infrastructure programme in order to ensure the transport of energy across the Union is a sine qua non of a working energy market. Thus, the importance of establishing and maintaining a fully integrated electricity and gas distribution grid can be observed within not only the internal dimension but also the external dimension. In the face of the unstable political environment of most of Europe's energy suppliers or transit countries, the security of supply has and will continue to have an impact on future price development. So far, Europe's economic malaise has dampened energy demand, but an integrated energy grid will be indispensable by the time the economy picks up again. Since building such a grid does not only cost enormous amounts of money but also takes considerable time to build, the time to invest is now. The benefits are huge. A complete internal energy market could bring net economic benefits between 16 and 40 billion euros a year (Booz & Company 2013). Thus, a well-integrated energy market enhances the resilience of the internal market and acts as a shock absorber for externally caused supply disruptions (European Commission 2014q). The only question that remains is who should pay for it?

Finally, even if one day in the future Europe achieves a fully integrated energy market, one further ingredient is required for it to be sustainable, and that is a low-carbon economy, which comes at a cost. Those costs are twofold. Despite the impressive progress energy efficiency has made over recent decades, sustainability cannot be achieved without a marked change in consumption patterns, because of negative externalities, irrespective of whether we produce energy from fossil fuels or renewables. All energy production is dirty. The best answer, therefore, is simply to use less. Investment into renewables, for which the EU has set an ambitious target of 27% of all energy production by 2030, is still a highly valuable objective, yet one that is difficult to achieve in times of economic crisis. Only in combination with massive investment in research and development, jointly organized at European level, can the full benefits of renewable energy sources be realized. Geographically, some countries are better positioned to harvest green energy. Meanwhile, the long-term installation of a Smart Grid will help cushion against the volatility of wind and solar power, and foster significant shares of cross-border trade.

An integrated market, based on a comprehensive legal framework and a flexible grid that allows for competition and transport of electricity and gas across the Union, is beneficial for the consumer and strengthens Europe's energy security. All of this can only be sustained through an evolved carbon-light economy, which amounts to nothing less than a fundamental transformation of European society.

Solidarity, common energy policy, or energy union?

At the beginning of Chapter 7, we noted that energy politics in the European Union is concomitantly an issue of integration and disintegration. Although energy matters (governance over the trade of coal) provided the impetus for Europe's integration in the 1950s, energy quickly became a central feature of Europe's national divisions rather than a source of unity. Almost seven decades later, a greatly expanded Union still struggles to agree on important details about a common energy policy. While the Lisbon Treaty pronounced the need for solidarity in the energy sector, it was less forthcoming on the details of how such solidarity could be achieved. Electricity trade between Spain and France provides an example. In 2014, Madrid complained that Paris was preventing the construction of transmission lines to carry Spanish electricity across the Pyrenees, while France complained that intermittent Spanish wind power posed risks to the French electricity grid. Under the notion of an internal energy market, the parties should be unified on the idea of moving electricity surpluses to lower costs for consumers. However, France has legitimate concerns about the stability of its largely nuclear-powered grid. In essence, France and Spain are electricity islands, a problem indicative of the fragmentation of Europe's energy market. Simply stated, harmonization seems difficult to deliver in the energy sector.

This is not to say that the member states disagree on everything, or even most things; a great deal of integration has already been achieved. Rather, it means that there is a fine and important difference between maintaining solidarity, either externally or internally, on energy issues, sharing targets and goals – and coordinating independent policies along such goals – and establishing a so-called Energy Union that establishes a single set of binding rules, targets, and coordinated purchasing groups executed through a single treaty-level authority. Therefore, Europe must choose which future it wants: independent states working together in harmony or a union that legally and practically speaks with one voice.

Cracks in EU solidarity are seen most vividly in the external dimension, but internally there are several key points of contention. At the March 2014 Council meeting it became evident that the member states remain divided over how much effort they are willing to invest to increase the 'greening' of their energy production. Some are investing billions in sustainability and renewables, while others continue to follow a simple market principle, one that favours coal over gas and renewables. When the Council met again in October, it endorsed the Commission's proposed collective targets 'in the most cost-effective manner possible [...] balancing considerations of fairness and solidarity' (European Council 2014). It also decided to reform the ETS, creating a stabilization fund, but agreed to indefinitely continue the system of free allocations, and decided on precious little in reference to the transportation sector. While the Council explicitly called the full implementation of the internal energy market a matter of urgency, it set only a few, low minimum standards on existing electricity interconnections and agreed to implement critical infrastructure projects, such as the North–South corridor, the Southern Gas Corridor, and a new gas hub in southern Europe. Moreover, despite noting its desire to build an Energy Union by keeping the goal 'under regular review', it repeated the traditional mantra that the EU must respect member-state freedom to determine their own energy mix. It would seem that regardless of its unity in goals, the particulars of who should do what, when, and how continues to elude the European Union.

Current discussions focus on a broad range of activities that would be part of an Energy Union. These include large-scale, joint R&D programmes to improve existing renewable technology, transparent systems for joint fossil resource management, gas storage, emergency distribution plans, authorized subsidies for gas in order to curb the increasing use of cheap coal, and the joint management of intermittency problems. None of these are particularly controversial and are unlikely to prove insurmountable obstacles. However, one notion most certainly is: the idea to implement a single buyer model for gas. While this would unite EU purchasing power, it would also make the Commission a central commodity manager that sets prices, bringing back the public utility model of the 1970s that the EU struggled so hard to liberalize. Whichever features it ultimately incorporates, such a Union could significantly improve relations with Russia and increase the EU's bargaining power in the medium to long term. Given the context of the 2014 Ukraine crisis and the obvious need to complete its internal energy market, an Energy Union for Europe seems less a matter of if than when. What remains are the form it takes and the competences allocated to its newly established authority, decisions that reside in the purview of diplomats.

Achieving climate targets and the future of renewables

The EU's ambitious targets, both for reducing its GHG emissions and increasing the share of renewables in the energy mix, are part of its plan to foster a sustainable, low-carbon economy. However, when the European Commission presented its framework for climate and energy policies for the period 2020–2030, in January 2014 (European Commission 2014), its proposals marked a tacit retreat on the strict environmental policies that have been the bulwark and driving force behind EU energy policy for decades. The Commission's proposal came at a time of increased concerns over European economic competitiveness. The EU was already lagging behind on its 20-20-20 targets, and there were real concerns that Europe was losing its economic edge, if not deindustrializing, as its environmental policies and increased energy costs were pushing energy-intensive industries to more attractive and less restrictive markets in the US and China. Their concerns were so substantial that the Commission proposed measures to drop binding targets for energy efficiency and keep a rather moderate share of renewables (27%), binding only at the EU level (Brutschin and Pollak 2014).

Most stakeholders agree that the EU's current policies to regulate its energy market are suboptimal, but that is where the agreement ends. Widespread discrepancy in terms of preferred goals for CO_2 emissions, energy efficiency, and the share of renewables remains. As described in Chapter 6, the EU will need to implement a number of structural reforms to account for massive surpluses in its ETS market, which have tanked carbon prices, if it is to achieve its ambitious climate goal of cutting CO_2 emissions by 40% from 1990 levels) by 2030. It will also need to come to terms with the possibility that it may not meet its renewable goals in its overall energy mix, let alone that of the individual member states.

The share of renewables in EU member states varies substantially, ranging from Sweden at 50% to the UK and the Netherlands, among others, relying on less than 10%. Variation in the share of renewables in the member states' energy mixes reveals an inherent dilemma: the wide gaps that exist between the member states call into question whether the EU's overall goals can be achieved, particularly in light of Germany's recent upswing in the use of cheap coal. Furthermore, the EU faces difficulty in setting nationally binding targets because to do so contradicts the principle of national sovereignty, and thus the EU is unlikely to move beyond collective targets even in the event that it creates an Energy Union. Thus, for the foreseeable future we can expect to see continued debate over the appropriate share of renewables in the EU and a series of non-disastrous crises resulting from missed targets.

While some countries (e.g. Germany, Spain, and the Netherlands) have invested heavily in renewable energy technology during recent years, creating thousands of new jobs and seeking high and binding targets in order to solidify the sector, others (e.g. the UK, Malta, and Luxembourg) would have to implement massive structural changes to come even close to the current target of 20% by 2020. Those that oppose renewable targets argue that they are the most expensive way to reduce carbon emissions, preferring instead, for example, to rely on gas as a transition fuel. Yet the pursuit of renewable targets is not economically inefficient per se, but rather, depends on geography and market conditions. Large-scale hydropower is cheaper than wind power. On the other hand, binding renewable targets provide an important signal for industry and could trigger future investments to induce structural changes in the energy market. Thus, the member states are unlikely to uniformly achieve overly ambitious targets for renewables, which puts pressure on them to fully integrate into the internal energy market, most notably in relation to a European-wide electricity grid, such that the achievements in renewables in one country can be explicitly felt in other more carbon-intensive countries.

In terms of energy efficiency, the Commission seems to have realized that setting absolute targets (e.g. maximum energy consumption of 1,474 Mtoe in 2020, as set down in 2012) does not make much sense given the strong influence of economic fluctuations on energy consumption (European Commission 2014: 22). One could as well reach such targets through another economic recession or miss them by a wide margin should the economy boom. Thus, the EU will need to focus on a target less sensitive to economic cycles and choose a better means of oversight in order to be more effective. Japan provides a good example. Tokyo was highly successful in increasing its energy efficiency by applying sector-specific goals. While such an approach may make oversight more difficult for Brussels, overseeing sector-specific energy efficiency achievements would provide the EU with a better control mechanism to respond to technological changes and economic cycles.

What all this means is that one has to be cautious about the outlook for the EU to achieve any, let alone all, of its lofty sustainability goals by 2030. Brussels simply cannot afford, or be expected, to trade economic competitiveness for environmental sustainability. It will need to continue tweaking its approach, part of which is underway with the planned reforms for the ETS. At the same time, the EU needs to balance the upside of energy-efficiency targets (i.e. its price signalling) with its downside (i.e. its dependence on economic performance). The first step in doing this is setting sector-specific targets. The most important, however, remains the establishment of a fully liberalized internal energy market. Unfortunately, achieving such a market seems to elude European politicians. Once again, the negotiations behind the 2030 framework revealed that the old game between the Commission and the member states continues unabated. The Commission proposes big changes, the Council confirms, but then waters them down, while the member states take their time selectively implementing those aspects that best befits their national interest. The EU's member states continue to sing the same song, but dance to a different tune.

Internal measures for infrastructure

In order to enhance security supply internally, the EU needs to be able move electricity and gas from any point within the Union to any other, based on demand. As former European Commissioner for Energy, Günther H. Oettinger, noted in 2014, with 'sufficient infrastructure, energy is produced where it's cheapest and sent to where it is needed. All this translates into secure energy supplies all over Europe and lower bills for consumers' (European Commission 2014k). While the EU made substantial progress in the electricity sector through the interlinking of grids and increased cooperation between national regulatory authorities, more investment is required in strategic cross-border infrastructure and the development of Smart Grids to allow for more small-scale generation and local supply security. Meanwhile, there is massive room for improvement in the gas sector.

Although many missing links connecting EU member states' gas networks have been completed or are under construction and reverse flow rules have been implemented, reducing individual member-state vulnerability to externally caused shortages, the central problem in the gas sector remains the lack of a compatible, high-volume, trans-European pipeline network connecting Europe's fledgling regional gas hubs. Europe will have to invest massively in its internal gas infrastructure to achieve this end. The question is then, who will pay for it? In accordance with its liberalization agenda, the general practice so far has been to allow the private sector to take the lead and, thus, the bulk of the risk. Due to the huge costs involved, however, few are willing, or able, to take on that risk, especially without getting exceptions from Brussels over whether they can both sell and transport gas. That leaves the member states, which until recently have been largely prohibited from biasing the market through selectively financing preferred national projects and companies. Thus, the EU must strike a new balance in this regard or implement a common pool for investments in trans-European gas networks.

The European Commission has already endorsed a list of 250 key energy infrastructure projects eligible for funding under the €5.85 billion Connecting Europe Facility (European Commission 2013q). These so-called projects of common interest (PCI) are earmarked for accelerated licensing and improved regulatory conditions, and of these, almost 100 projects are in the field of gas transmission, storage, and LNG. However, while this may sound like a large amount of money, it amounts to just a fraction of the estimated cost. By comparison, the US, which is not known for over-intervention into its energy market, spent more than $90 billion on clean energy investments through its 2009 stimulus package. The Commission is thus literally banking on the assumption that initial seed money and regulatory easing will leverage the necessary private and public funding for these projects. In so doing, it is taking little risk and expecting huge results. The EU needs to spend a great deal more money collectively on its energy infrastructure, particularly in the gas sector. It needs to invest in large-scale regional gas hubs and storage facilities, dually enhancing security of supply and creating an internal mechanism for price signalling. While there seems no shortage of efforts to promote external projects to bring external gas to Europe, efforts at home need to be increased in order to capitalize on the changing nature of the international gas market.

Increased indigenous supplies: The 'hype' and reality of fracking in the EU

Europe's indigenous fossil fuel production has been declining for decades, pitting the need for long-term security of supply against the reality of having to replace conventional oil and gas supplies with a substantial reduction in consumption or an increase in unconventional sources. Notwithstanding increases in the development of Arctic resources and the discovery of a large gas field off the shore of Cyprus, there is an increasing interest in recovering unconventional fossil resources; and the shale gas revolution in the US has heightened speculation over the potential for shale gas to similarly transform energy markets in Europe, where early-stage exploration is under way. The European Commission estimates that the addition of unconventional gas into the EU energy mix could help alleviate import dependence in some member states and meet up to 10 % of EU gas demand by 2035, contributing to both EU security of supply and competitiveness. Yet of the technically recoverable shale gas resources estimated to be just over 13,000 Mtoe, only a fraction are considered to be economically recoverable (European Commission 2014l) and the EU will have to find ways to reconcile developments in shale gas exploitation with its environmental stewardship. For better or worse, the revolution underway in the US is not likely to be replicated in Europe.

As in many other policy areas, the member states and European citizens are deeply divided over unconventional gas, particularly because of the method used to extract it. Due to the potential environmental risks, fracking poses serious problems for social acceptance. As one can imagine, this has created quite some controversy, as Europeans are deeply divided over both its potential and environmental impact. A public consultation conducted by the Commission in 2012 elicited more than 23,000 responses, revealing substantial opposition from countries such as France on one hand and cautious support from populations in Poland and the UK on the other. In 2011, France became the first EU member state to ban fracking altogether, later converting that into a temporary ban, and since then the UK and Poland have moved forward on exploration, albeit rather unsuccessfully. As expected, the Commission has taken a neutral position on fracking, in line with the mantra that the member states alone are responsible for determining their energy mix. It did issue a recommendation in January 2014 on minimum principles, using high-volume hydraulic fracturing, and has since established a network of experts on the subject to advise it, but it has so far not taken a formal position on the matter other than to pigeonhole it in similar fashion as it does nuclear power; it simply is not their bailiwick.

Beyond the obvious environmental concerns, there also are serious practical obstacles to successfully tapping unconventional fossil resources in Europe, and even marginally replicating the American experience. The first problem is the permit process. Among the EU member states, land ownership does not automatically include ownership of the resources beneath the surface. Such resources are generally owned by the State, unless legally specified otherwise, which hinders private investment and opportunity, and thus, also reduces incentives for the population to support exploration. The next problem is economic viability. Unconventional drilling in Poland, for example, cost as much as three times that in the US according to Schlumberger, the world's largest oilfield-services provider (Strzelecki 2011). Meanwhile, even if European companies could overcome the expenses and regulatory bottlenecks, Europe's oil and gas field sectors lack much of the suitable equipment and experience to exploit what resources might be there. According to a 2013 Ernst and Young report, Europe actively operates fewer than 50 onshore exploration rigs at any one time. By comparison, more than 2000 are in operation in the

US (EYMG 2013). Finally, in the most lucrative markets and locations where unconventional gas resources are most likely to be exploitable – namely, the UK, France, and Germany – the locations of fields are either too close to the underlying water table or located in densely populated areas, rendering exploration itself a political nightmare.

Despite all these obstacles and no sign that the EU is moving towards a unified position on fracking (Poland and the UK have already opposed an EU-wide ban), exploration for unconventional gas is moving forward in Europe. Deposits in the UK and Poland, as well as the Baltic basin, the Pannonian–Transylvanian basin in Hungary and Romania, and the Carpathian–Balkanian basin of Romania and Bulgaria (KPMG 2012), have all been identified as potentially rich sources of unconventional gas. Their addition to the EU's energy mix will hardly rescue the EU from its overwhelming dependence on foreign suppliers; although it could alleviate some of the pressure on its Central and East European members in the medium to long term. Nevertheless, shale gas is unlikely to change the overall energy picture in Europe. If it has any impact at all in the near future, it is more likely to follow an evolutionary rather than revolutionary path (EYMG 2013). Therefore, Europe's best option remains a combination of reduced consumption, greater use of renewables, the transformation of its transport and heating sectors to electricity, and above all else, securing the reliable and affordable delivery of external fossil fuel supplies.

The EU's geopolitical challenges

The case of Russia

Russian President Vladimir Putin's vision of a resurgent Russia is tied to state control of energy resources at home and abroad. It is a policy approach directly at odds with the EU's open-market ideals. In the late 1990s, Mr Putin wrote that state control of mineral resources should form the basis for Russia's re-emergence as a global economic and political power, and that such government stewardship of the extractive industries would provide the country with the ability to 'have a large impact on the world commodities market' (Balzer 2006: 51). Although Putin's academic writings predate his career as a politician, his framework for Russian reconstruction has been fulfilled in correlation with his earlier study.

Once in office, he oversaw the renationalization of Russian oil and gas resources and, under his tenure, Russia and Gazprom worked diligently to close bilateral agreements with individual European member states and companies, at the cost of European unity. Seen in the light of Putin's grand plan, Russia's concerns about EU diplomatic efforts to export common rules rooted in the EU's energy acquis, its Neighbourhood Policy, its seemingly unstoppable enlargement process, its attempts to meddle in the political affairs of Ukraine and Georgia, and its efforts to bypass Russia in accessing Caspian Littoral resources, take on added meaning. These areas are not only historically central to Russian security thinking but also an essential part of the country's national energy-export infrastructure, and thus, an indispensable component of Russian capability to project economic power.

East and south-east Europe constitute both the EU's contemporary energy lifeline and its main choke point, and Russia knows this all too well. Most Russian energy exports to Europe flow either through Belarus or Ukraine, although substantial amounts now flow directly from Russia to Germany via the Nord Stream pipeline. Price disputes between Ukraine and Russia led to several brief, but painful, natural gas cut-offs during the coldest winter months between late 2005 and early 2009. A similar dispute with Belarus led to a halt in oil supplies in 2007. Russia's intervention in Georgia in 2008 led to a temporary stoppage of oil flows (from Azerbaijan to Turkey) through that country, and when it seized the Crimea in 2014, the largest importer of Russian gas, Germany, did little more than offer tough talk. As Donald Tusk, the prime minister of Poland, tersely noted at the time, 'Germany's reliance on Russian gas can effectively limit European sovereignty' (Smale 2014). For Moscow and its strategy to keep its gas flowing to Europe and the latter's money flowing back to Moscow, Ukraine's vacillating orientation between East and West is unacceptable.

Moscow's concerns are not unwarranted. The break-up of Yugoslavia, long an ally of Russia, pitted the country against NATO. Later efforts to bestow Kosovo with independence, under UN Security Council Resolution 1244, only heightened concerns of EU and NATO expansion. For Moscow, the move set a precedent for ethnic separatist movements all along the Russian periphery, a concern seconded by Oxford University's Timothy Garton Ash, who, in 2008, noted, 'there will be more Kosovos' (Ash 2008). Indeed, Russian President Dimitry Medvedev alluded to as much in an

August 2008 interview with the BBC (BBC 2008a), only two weeks after calling Russia the region's 'guarantor of peace' (BBC 2008b). Furthermore, Russia is concerned about the spill over effects caused by domestic instability in Ukraine and Belarus, and with good reason. Its southern borders are porous at best, and the 2004 Orange Revolution and 2014 ousting of pro-Russian President Yanukovych in Ukraine reminded Moscow of the vulnerable nature of the countries along its periphery.

Moscow considers most of Eastern Europe, including several states that joined the EU, as pivotal to Russian national heritage and security. Both Ukraine and Belarus have strong cultural ties to Russia, including widespread use of the Russian language (Oliker et al. 2009). The EU's efforts to negotiate an Association Agreement with Kiev directly challenged Russia's so-called sphere of influence and exacerbated Russian concerns about EU intentions. Given the constellation of powers and interests in east and south-east Europe, there should be little wonder why Russia is seeking alternative energy transit routes around both Ukraine and Belarus. Since the region also serves as a possible future transit route for Caspian and Central Eurasian energy resources destined for the EU, Europe can expect heightened tensions with Russia over everything from pipeline routes to election monitoring.

For the EU to secure its short- and medium-term energy needs, it may need to placate Russian concerns and cater to Moscow's demands for regional domination and neighbourly suzerainty. Doing this, however, runs contrary to Brussels' political leanings and interests. Clearly, Europe needs a strategy to deal with its massive eastern neighbour and energy supplier. Public declarations to reduce dependency on Russia are not going to convince the Russians to loosen their grip on the Eastern European transit states. In fact, it may cause the opposite effect, with the additional detriment of dividing the Union.

The race for Caspian resources

The Caspian Littoral states (Russia, Iran, Kazakhstan, Turkmenistan, and Azerbaijan) and the South Asian countries of Tajikistan and Kyrgyzstan are rich in oil and natural gas supplies. Three of the five Caspian countries are former members of the Soviet Union, as are their two eastern neighbours and as far as Moscow is concerned, they are as much part of Russia's sphere of influence as the countries of Eastern Europe. Russia operates military bases in Tajikistan, Kyrgyzstan, and Kazakhstan, the latter of which hosts Russia's Baikonur Cosmodrome from where Moscow still operates its space-launch facilities. This region was the hotbed of a century-long geopolitical struggle for supremacy between the British Empire and Czarist Russia, stretching from Iran to China (Allworth 1994, Hopkirk 1994). Although Great Britain no longer plays a major role in the region, the game to control Central Eurasia still rages.

Most of the region's resources (drawn in 2012 from proven reserves of circa 40 billion barrels of oil and 28 trillion cubic metres of gas) are exported via pipelines operated by Russian state-owned companies. Independence from Moscow, however, has opened the doors to European, Chinese, and Indian energy interests. Awash in cash from higher energy prices, the governments of Central Eurasia are increasingly toying with the idea of letting the various competitors outbid each other for access.

Europe faces more competition for energy resources in Central Eurasia than anywhere else in the world. Both China and Russia straddle the region. The EU does not. Therefore, in order to secure the region's oil or gas, either Brussels or a collection of EU-based companies, will have to spend a fortune, and do so almost entirely on local terms. Even if Europe were ultimately to succeed in securing contracts and building pipelines, it faces the additional dilemma that the exporting countries may not be able to meet their commitments. Russia and China already have secured long-term purchase contracts for the region's gas, and Beijing is investing in new pipeline networks, while Russia is working to upgrade it existing lines.

There are many security issues in Central Asia. The potential for conflict, if not chaos, cannot be ignored. Russia's long southern border with the region provides multiple points of entry into the country for the smuggling of people, weapons, and illegal drugs, all of which are connected by land routes to Europe. Chechnya is not far away and has been the irritable focus of Russian military adventures since the late eighteenth century, when Czarist Russia took control of the territory from the Turkish and Persian Empires. Moreover, authoritarian leaders that oppress opposition movements and use energy revenues to shore up their military capacities govern many of the region's potential energy suppliers. The region's politics were only complicated by events in Afghanistan. Consequently, the need for reliable supply lines and the ability to launch emergency search-and-rescue operations provided an impetus for the construction of American military bases scattered throughout the region. As in the case of Ukraine and Georgia, the presence of outside military forces in the region is a challenge to Russian dominance, despite Moscow's initial approval.

The EU is at a distinct disadvantage in Central Eurasia, where Russia is without question the region's dominant military and economic power, and has demonstrated its capability and will to secure its interests across the region repeatedly over the last two centuries. Moscow also remains the primary trading partner for all the former Soviet Republics, although China is encroaching upon Russian economic interests in the region (Sutter 2012). Beijing is clearly capable of securing their interests with money if necessary, whereas Europe appears to be unprepared, and particularly slow, to act on either account. Moreover, Russia, China, and India do not demand domestic political reforms in exchange for their business, a point that strengthens their negotiating position and weakens EU's effort to export its energy acquis.

Meanwhile, the United States is no less interested in the region's energy resources. Although Central Asia is very unlikely to ever supply US markets, boosting regional energy exports free from Russian influence does appear to be part of its two-pronged strategy in the region, to both contain the spread of radicalism emanating from Afghanistan, Pakistan, and Iran on one hand and limiting the spread of Russia's neo-mercantilist approach to international politics on the other (Oliker and Schlapak 2005). Combined with aggressive efforts by China and Russia to secure control over the region's energy resources, the US and EU presence is complicating the delicate balance of power in the region.

Defining the relationship with Turkey

Unlike Ukraine, Belarus, or Georgia, Turkey, which sits at the crossroads between Europe and the resource-rich regions of the Middle East and South-Central Asia, is deeply embedded in the process of negotiating accession to the EU. An associate member of the EU since 1962 (it signed a Customs Union agreement with Brussels in 1995), it has been an official candidate for membership since 1999 and began formal negotiations for accession in 2005. Turkey joined NATO in 1952, and for much of the Cold War, was Europe's southern line of defence. It controls a vast swathe of the Black Sea basin, and borders Georgia, Armenia, Azerbaijan, Iran, Iraq, and Syria. It has ports on the Mediterranean, Aegean, Black, and Marmara Seas, and it is on the receiving end of pipelines that run from Iran and Iraq. Simply stated, Turkey plays a central role in Europe's strategy to reduce its energy dependence on Russia.

Turkey's role as a transit state in the originally planned Nabucco project, its arrangements with Russia to reroute the South Stream pipeline through to Turkey to avoid crossing Ukraine, and its independent plans to build the Trans-Anatolian Pipeline (TANAP) with Azerbaijan, all point to the central role that Turkey plays in the EU's energy future. TANAP is set to become the backbone of the EU's new Southern gas corridor, supposedly capable of delivering up to 40 billion cubic metres of gas annually for Turkey's western border to key European hubs: in Hungary via the South-East European Pipeline (SEEP), Austria via Nabucco West, and Greece and Italy via the Trans-Adriatic Pipeline (TAP). In welcoming the finalization of the agreement and authorization to begin construction of TANAP, EU Energy Commissioner Günther Oettinger welcomed the Turkish–Azeri project as one that brings the EU 'a step closer to its aim to get gas directly from Azerbaijan and the other countries in the Caspian region' (European Commission 2012h). Yet like so many pipeline projects, there are serious reasons to question such exuberant optimism. Once the gas arrives in western Turkey the question becomes, how do we bring it into the European network? Moreover, when TANAP finally comes online, it will carry much less gas than its backers claim (only 16 bcm, or roughly half of what Nabucco originally promised), and of that, 6 bcm will stay in Turkey, leaving only 10 bcm destined to Europe via TAP and/or others (Belkin et al. 2013, ENTSOG 2013a: 70). Hypothetically, Turkish Stream will further congest the network, adding upwards to 50 bcm, which suggests that regardless of what comes of any of these projects, the EU should be investing heavily in the South-Eastern Corridor's transport capacity.

Officially, Turkey wants to join the EU, albeit its increasingly eastern orientation has become somewhat visible in the past decade. EU member states, however, are divided over Turkish accession, and the arduous process to satisfy all of the EU's demands have left some Turks feeling jilted (Gulmeza 2013, Morelli 2013, Wood 2013). Even if Ankara were to succeed in adopting all of the standards required for accession, it remains unclear as to whether any of the current EU members will ultimately veto Turkey's accession. If Turkey does ultimately become a member of the EU, it will not solve the latter's energy problems, but it will bring the evolving internal energy market much closer to vast sources of oil and gas right across its south-east border. If it does not join, the EU will have to find a way to avoid a messy divorce. There is always the danger that instead of creating a friendly energy-transit partner, an EU–Turkey break-up could initiate the rise of a profiteering regional competitor. There is good money to be made in delivering gas to the EU, and the profits Turkey could reap from the purchase and resale of regional gas could be an unwelcome, but necessary price of settlement. As a member or not, Europe needs Turkey if it wants unfettered (i.e. not Russian) access to the energy resources of the Middle East and Caspian Littoral states.

By initiating decades-long negotiations with Turkey and by locking the EU into plans to access the resources east of Turkey, Brussels has entangled itself into a geopolitical morass of which there is no easy way out. Turkey is literally Europe's bridge to Asia. Ankara plays an essential security role across the entire Eastern Mediterranean. It is the only real challenger to Russian domination of the Black Sea, and a core intermediary for the Arab–Israeli conflict. These are all issues of essential value to Europe's long-term physical and economic security.

Responsibly exploiting the Arctic

Europe's brightest hope for long-term security of supply may be nearby, just above the Arctic Circle, a vast and uninviting region that encompasses about 6% of the Earth's surface. The United States Geological Survey (USGS) estimates that the Arctic contains as much as 25% of the world's undiscovered fossil fuel resources, including 90 billion barrels of oil and 48 trillion cubic metres of natural gas, 84% of which are to be found under the region's disputed international waters (USGS 2008). If only a quarter of that gas were delivered to Europe, it could secure Europe's fossil fuel needs well into the twenty-second century.

Under debate in the Arctic are both the environmental sustainability of its prospective development and the final delineation of international borders between the circumpolar states. The outcome of the latter is important to the EU because it will determine the region's exclusive economic zones. The 1982 United Nations

Convention on the Law of the Seas (United Nations 1982), grants coastal countries control of resources on and beneath the ocean floor beyond the limits of its national jurisdiction to a distance of 200 nautical miles (370.4 km). If USGS estimates prove even close to correct and Brussels fails to secure Europe's territorial claims, or at least hedge against Russia's, EU consumers may be forced to rely on proxies for future access to the region's vast economic potential. Canada, Denmark, Norway, the US, and Russia are all engaged in Arctic territorial disputes, and both Russia and Canada lay claim to the region's petroleum. Safely accessing the resources in the region will require a high level of international political, commercial, and scientific cooperation. For the EU, it is an opportunity to promote multilateral, if not even global, governance on the exploration and exploitation of undiscovered energy resources.

The EU has already funded several projects, prior to 2008, to prepare for Arctic development, including the Arctic Operational Platform Project and the ICE ROUTES, an ice meteorology study of the North Sea (ECS 2008: 188). Since then, it has stepped up its activities considerably, investing over €1 billion 'to develop the economic, social and environmental potential of the Arctic regions of the EU and neighbouring areas', and between 2007 and 2013, contributed circa €200 million to international research activities in the region (through the Seventh Framework Programme). In 2012, the Commission and the High Representative of the Union for Foreign Affairs and Security Policy, Catherine Ashton, issued a joint Communication, *Developing a European Union Policy towards the Arctic Region: Progress since 2008 and Next Steps* (European Commission 2012c), in which they spelt out EU strategy and concerns in the area. Among these were substantially increased levels of cooperation with the international and multilateral organizations, such as the Arctic Council and the Barents Euro-Arctic Council, of which the Commission is a member. The Commission is also a member of the so-called Northern Dimension, where it works together with Iceland, Norway, and the Russian Federation.

Exploiting the Arctic will offer both opportunities and risks for the EU. Given the stakes in terms of security of supply, possible environmental damage to the Arctic's pristine waters, and the political standing of the EU vis-à-vis its neighbours, EU Arctic policy can be expected to play an increasingly important role in the external dimension of EU energy policy.

Concluding remarks

The EU member states broadly agree on the need for comprehensive energy security, but there remains a lot of unfinished business (both internally and externally) to be resolved. It must bring its internal energy market to fruition, determine which measures are best to improve efficiency and climate change goals, and move forward on building an Energy Union. In order to create a carbon-light economy, it will have to not only evaluate the economic viability of renewables in the long term but also explore the possibility, and recognize the limitations, of increasing indigenous fossil fuel exploration. Along the way, it must also face and resolve a series of internal and external challenges, some of which we have spelt out. Others, as yet unknown of, are certain to appear.

Meanwhile, planned projects around its periphery could greatly reduce EU dependence on Russian gas by 2030. The international LNG market is growing. The Southern Corridor and access to Caspian gas looks to become a reality by 2020. In the process of extending its pipeline reach and increasing the number of routes, the EU is in some cases advertently, and in others inadvertently, challenging the vested interests of some of its neighbours. Smoothing relations with Russia and Turkey, finding the most cost-effective way to access Caspian resources, and determining how to exploit the vast resources of the Arctic without destroying its pristine nature are just some of the most pressing issues the EU faces.

Yet the future is hard to see; just as the fictional Jedi master, Yoda, observed, 'Always in motion is the future'. In terms of energy politics, this future will be a mixture between the further development of 'green energy', the diversification of suppliers, routes, and sources, technological progress, a much more responsible use of energy, and diplomatic prudence. However, fossil fuels will continue to occupy a central position in the near future. Reshaping the technological foundations of our fossil fuel-based societies will take a long time. As impressive as the advance of green energy technology is, it is not a panacea for all our energy needs. Only time can tell whether EU solidarity and technological innovation will withstand the pressures of national interest, bilateral relations, and volatile demand.

References

AEA. (2006), *General Comments on the Working Document of the Rapporteur Caroline Lucas on Reducing the Climate Change Impact of Aviation*, Position Paper, 13 April 2006.

Agency for the Cooperation of Energy Regulators (ACER). (2014), *The EU Energy Market*, http://www.acer.europa.eu/the_eu_energy_market/Pages/default.aspx [30 April 2014].

Allworth, E.E. (Ed.). (1994), *Central Asia: 130 Years of Russian Dominance*, Durham: Duke University Press.

Andersen, S.S. (2000), *EU Energy Policy. Interest Interaction and Supranational Authority*, Arena Working Paper No. 5.

Ash, T.G. (2008), 'The Kosovo Precedent' [21 February 2008], *The Los Angeles Times,* http://www.latimes.com/news/la-oe-ash21feb21,0,2808437.story#axzz2whdWHXNI [23 March 2014].

Balzer, H. (2006), 'Vladimir Putin's Academic Writings and Russian Natural Resource Policy', *Problems of Post-Communism,* 53(1), 48–54.

Baumgartner, F.R. and Jones, B.D. (1991), 'Agenda Dynamics and Policy Subsystems', *Journal of Politics*, 53(4), 1044–1074.

BBC. (2008a), *Russia Recognises Georgian Rebels* [26 August 2008], http://news.bbc.co.uk/2/hi/europe/7582181.stm [23 March 2014].

——— (2008b), *Russia 'Guarantor' of Peace* [15 August 2008], http://news.bbc.co.uk/2/hi/europe/7563624.stm [23 March 2014].

Belkin, P., Nichol, J., Ratner, M. and Woehrel, S. (2013), *Europe's Energy Security: Options and Challenges to Natural Gas Supply Diversification*, Washington, DC: Congressional Research Service.

Berg, W. (Ed.). (2001), *Regulierte Selbstregulierung als Steuerungskonzept des Gewährleistungsstaates*, Berlin: Duncker & Humboldt.

Beyli. (2012), *EU External Energy Policies*, *POLINARES* Working Paper No. 69, Centre for Energy, Petroleum and Mineral Law and Policy (CEPMLP) at the University of Dundee, http://www.polinares.eu/docs/d5-1/polinares_wp5_chapter3.pdf [20 March 2014].

Bialek, J.W. (2004), *Recent Blackouts in US and Continental Europe: Is Liberalisation to Blame?*, Cambridge Working Papers in Economics CWPE 0407 / CMI Working Paper 34, Cambridge: University of Cambridge.

Birchfield, V.L. and Duffield, J.S. (Eds.). (2011), *Toward a Common European Union Energy Policy. Problems, Progress, and Prospects,* New York: Palgrave Macmillan.

Bosseboeuf, D., Lapillonne, B., Pollier, K. and Sebi, C. (2013), *Energy Efficiency Trends in the EU, Lessons from the Odyssée-Mure Project*, Paris: ADEME.

Booz & Company. (2013), *Benefits of an Integrated European Energy Market, Prepared for Directorate-General Energy European Commission,* Amsterdam: Booz & Company.

BP. (1987), *Statistical Review of World Energy,* London: BP.

——— (2007), *Statistical Review of World Energy,* London: BP.

——— (2013), *Statistical Review of World Energy,* London: BP.

——— (2014), *BP Energy Outlook 2035,* London: BP.

Breakthrough Institute. (2011), *Energy Emergence. Rebound and Backfire as Emergent Phenomena,* http://thebreakthrough.org/blog/Energy_Emergence.pdf [18 March 2014].

Brutschin, E. (2013), 'Dynamics in EU Policy-Making: The Liberalization of the European Gas Market', Unpublished Dissertation, University of Konstanz, http://nbn-resolving.de/urn:nbn:de:bsz:352-253135 [30 April 2015].

Brutschin, E. and Pollak, J. (2014), *The European Commission 2030 Framework for Climate and Energy Policies: A Step Back?*, Standpunkt: Position 21/2014, Vienna: Institute for Advanced Studies, https://www.ihs.ac.at/fileadmin/public/user_upload/Standpunkt_21_2030_energy-policy.pdf [20 October 2014].

Burger, B. (2013), *Electricity Production from Solar and Wind in Germany in 2013*, Freiburg: Fraunhofer Institute for Solar Energy Systems ISE.

Cardis, E., et al. (2006), 'Cancer Consequences of the Chernobyl Accident: 20 Years On', *Journal of Radiological Protection*, 26(2), 127–140.

Christiansen, T., Jorgensen, K.E. and Wiener, A. (1999), 'The Social Construction of Europe', *Journal of European Public Policy*, 6(4), 528–544.

Christoffersen, P.S. (2009), 'The role of the European Council in institutional and policy developments in the European Union', in: Phinnemore, D. and Warleigh-Lack, A. (Eds.), *Reflections on European Integration. 50 Years of the Treaty of Rome*, Houndmills: Palgrave Macmillan, pp. 74–92.

Ciuta, F. (2010), 'Conceptual Notes on Energy Security: Total or Banal Security?', *Security Dialogue*, 41(2), 123–144.

CNN. (2008), 'Interview with Larry King' [24 September 2008], *Cable News Network*, http://transcripts.cnn.com/TRANSCRIPTS/0809/24/lkl.01.html [30 April 2014].

Council of the European Communities. (1968), *Council Directive of 20 December 1968 Imposing an Obligation on Member States of the EEC to Maintain Minimum Stocks of Crude Oil and/or Petroleum Products*, 68/414/EEC [OJ 1968, L 308/14].

——— (1974), *Council Resolution of 17 December 1974 Concerning Community Energy Policy Objectives for 1985* [OJ 1975, C 153].

——— (1975), *Council Directive of 20 May 1975 Obliging the Member States to Maintain Minimum Stocks of Fossil Fuel at Thermal Power Stations*, 75/339/EEC [OJ 1975 L 153/35].

——— (1976a), *Council Directive of 4 May 1976 Regarding a Community Procedure for Information and Consultation on the Prices of Crude Oil and Petroleum Products in the Community*, 76/491/EEC [OJ 1976, L 140/4].

——— (1976b), *Council Regulation of 21 June 1976 Concerning the Communication of Information on the State of the Community's Energy Supplies*, 1729/76/EEC [OJ 1976, L 198/1].

——— (1977a), *Council Decision of 14 February 1977 on the Exporting of Crude Oil and Petroleum Products from One Member State to Another in the Event of Supply Difficulties*, 77/186/EEC [OJ 1977, L 61/23].

——— (1977b), *Council Decision of 7 November 1977 on the Setting of a Community Target for a Reduction in the Consumption of Primary Sources of Energy in the Event of Difficulties in the Supply of Crude Oil and Petroleum Products*, 77/706/EEC [OJ 1977, L 292].

——— (1980), *Resolution Embodying the Opinion of the European Parliament on the Communication from the Commission of the European Communities to the Council on the Energy Objectives of the Community for 1990 and Convergence of Policies of the Member States on Nuclear Energy and Energy Policy* [OJ 1980, C 59/41].

——— (1986), *Council Resolution of 16 September 1986 Concerning New Community Energy Policy Objectives for 1995 and Convergence of the Policies of the Member States*, 86/C241/01 [OJ 1986, C 241/1].

——— (1987), *Council Decision of 14 December 1987 on Community Arrangements for the Early Exchange of Information in the Event of a Radiological Emergency*, 87/600/Euratom [OJ 1987, L 371/76].

——— (1987a), *Council Regulation of 22 December 1987 Laying Down Maximum Permitted Levels of Radioactive Contamination of Foodstuffs and of Feeding Stuffs Following a Nuclear Accident or Any Other Case of Radiological Emergency*, 3954/87/Euratom [OJ 1987, L 371/11].

——— (1989), *Council Directive of 27 November 1989 on Informing the General Public About Health Protection Measures to be Applied and Steps to be Taken in the Event of a Radiological Emergency*, 89/618/Euratom [OJ 1989, L 357/31].

——— (1990a), *Council Directive 90/377/EEC of 29 June 1990 Concerning a Community Procedure to Improve the Transparency of Gas and Electricity Prices Charged to Industrial End-Users* [OJ 1990, L 185/16].

——— (1990b), *Council Regulation of 7 May 1990 on the Establishment of the European Environment Agency and the European Environment Information and Observation Network*, 1210/90/EEC [OJ 1990, L 120/1].

——— (1990c), *Council Directive 90/547/EEC of 29 October 1990 on the Transit of Electricity through Transmission Grids* [OJ 1990, L 313/30].

——— (1990d), *Council Directive of 4 December 1990 on the Operational Protection of Outside Workers Exposed to the Risk of Ionizing Radiation during their Activities in Controlled Areas*, 90/641/Euratom [OJ 1990, L 349/21].

——— (1991), *Council Directive 91/296/EEC of 31 May 1991 on the Transit of Natural Gas through Grids* [OJ 1991, L 147/37].

Council of European Energy Regulators (CEER). (2014), 'Homepage of the European Energy Regulators' http://www.ceer.eu/portal/page/portal/EER_HOME [30 April 2014].

Council (of the European Union). (1996), *Council Directive 96/29/EURATOM of 13 May 1996 Laying Down Basic Safety Standards for the Protection of the Health of Workers and the General Public against the Dangers Arising from Ionizing Radiation* [OJ 1996, L 159/1].

——— (1997), *Council Directive 97/43/Euratom of 30 June 1997 on the Health Protection of Individuals against the Dangers of Ionizing Radiation in Relation to Medical Exposure, and Repealing Directive 84/466/Euratom* [OJ 1997, L180/22].

——— (2001), *Report on Nuclear Safety in the Context of Enlargement (9181/01)*, http://www.eumonitor.nl/9353000/1/j4nvgs5kjg27kof_j9vvik7m1c3gyxp/vi7jgsr87wzu/f=/9181_01_add_1.pdf [30 April 2015].

——— (2002), *COREPER 'I' Item Note: Peer Review Status Report (9601/02)*, Brussels, 5 June 2002, https://www.sujb.cz/fileadmin/sujb/docs/zpravy/ostatni_zpravy/aqg.pdf [30 April 2015].

——— (2002a), *Council Decision of 25 April 2002 Concerning the Approval, On Behalf of the European Community, of the Kyoto Protocol to the United Nations Framework Convention on Climate Change and the Joint Fulfilment of Commitments Thereunder* (2002/358/CE) [OJ 2002, L 130/1].

——— (2003), *Council Directive 2003/122/Euratom of 22 December 2003 on the Control of High-Activity Sealed Radioactive Sources and Orphan Sources* [OJ 2003, L 346/57].

——— (2003a), *Proposal for a Council (Euratom) Directive Setting Out Basic Obligations and General Principles on the Safety of Nuclear Installations COM, (2003) 32 Final,* http://eur-lex.europa.eu/legal-content/EN/TXT/PDF/?uri=CELEX:52003PC0032%2801%29&qid=1432467144298&from=EN [30 April 2014].

——— (2003b), *Council Regulation (EC) No 1/2003 of 16 December 2002 on the Implementation of the Rules on Competition Laid Down in Articles 81 and 82 of the Treaty* [OJ 2003, L 001/1].

——— (2004), *Commission Recommendation of 18 December 2003 on Standardised Information on Radioactive Airborne and Liquid Discharges into the Environment from Nuclear Power Reactors and Reprocessing Plants in Normal Operation (Notified under Document Number C(2003) 4832)* [OJ 2004, L 002/36].

——— (2006), *Council Directive 2006/117/Euratom of 20 November 2006 on the Supervision and Control of Shipments of Radioactive Waste and Spent Fuel* [OJ 2006, L 337/21].

——— (2006a), *Proposal for a European Energy Security Treaty – Presentation by the Polish Delegation (Poland's Non-Paper: Outline of the European Energy Security Treaty) 7160/06*, Brussels [9 March 2006], http://register.consilium.europa.eu/pdf/en/06/st07/st07160.en06.pdf [30 April 2015].

——— (2006b), *Council Decision of 29 May 2006 on the Conclusion by the European Community of the Energy Community Treaty* (2006/500/EC) [OJ 2006, L 198/15].

——— (2006c), *The Energy Community Treaty* [OJ 2006, L 198/18].

——— (2009), *Joint Declaration of the Prague Eastern Partnership Summit*, Brussels, 7 May 2009, 8435/09 (Presse 78), Prague, http://www.consilium.europa.eu/uedocs/cms_data/docs/pressdata/en/er/107589.pdf [30 May 2015].

——— (2009a), *Council Directive 2009/71/EURATOM of 25 June 2009 Establishing a Community Framework for the Nuclear Safety of Nuclear Installations* [OJ 2009, L 172/18].

——— (2009b), *Council Decision of 1 December 2009 Adopting the Council's Rules of Procedure* (2009/937/EU) [OJ 2009, L 325/35].

——— (2010), *Joint Statement on the Partnership for Modernisation EU-Russia Summit 31 May–1 June 2010*, Rostov-on-Don, 1 June 2010, 10546/10 (Presse 154), http://europa.eu/rapid/press-release_PRES-10-154_en.htm [30 April 2015].

——— (2011), *Council Conclusions on Strengthening the External Dimension of the EU Energy Policy, 3127th TRANSPORT, TELECOMMUNICATIONS and ENERGY Council Meeting (Energy Items)*, Brussels, 24 November 2011, http://www.consilium.europa.eu/uedocs/cms_data/docs/pressdata/en/trans/126327.pdf [30 April 2015].

——— (2013), *Follow-up to the European Council of 22 May 2013: Review of Developments on the External Dimension of the EU Energy Policy* (17756/13), http://register.consilium.europa.eu/doc/srv?l=EN&f=ST%2017756%202013%20INIT [30 April 2015].

——— (2013a), *Council Directive 2013/59/Euratom of 5 December 2013 Laying Down Basic Safety Standards for Protection against the Dangers Arising from Exposure to Ionising Radiation, and Repealing Directives 89/618/Euratom, 90/641/Euratom, 96/29/Euratom, 97/43/Euratom and 2003/122/Euratom* [OJ 214, L 13/1].

——— (2014), *Note from the General Secretariat to Delegations: List of Council Preparatory Bodies*, Subject: POLGEN 7, Document No. 12111/1/13, January 2010. Public Register of Council Documents, http://www.palkeet.fi/openfile/98 [30 April 2015].

——— (2014a), *Council Directive 2014/87/Euratom of 8 July 2014 Amending Directive 2009/71/Euratom Establishing a Community Framework for the Nuclear Safety of Nuclear Installations* [OJ 2014, L 219/42].

——— (2015), *Council of the European Union, Configurations: Transport, Telecommunications and Energy Council configuration (TTE)*, http://www.consilium.europa.eu/en/council-eu/configurations/tte/ [30 April 2015].

Council (of the European Union) and the Commission (of the European Communities). (1997), *Council and Commission Decision of 30 October 1997 on the Conclusion of the Partnership and Cooperation Agreement between the European Communities and their Member States, of the One Part, and the Russian Federation, of the Other Part (97/800/EC, ECSC, Euratom)* [OJ 1997, L 327/1].

——— (1998), *98/181/EC, ECSC, Euratom: Council and Commission Decision of 23 September 1997 on the Conclusion, by the European Communities, of the Energy Charter Treaty and the Energy Charter Protocol on Energy Efficiency and Related Environmental Aspects* [0J 1998, L 69/1-116].

——— (1998a), *Council and Commission Decision of 28 May 1998 on the Conclusion of the Partnership and Cooperation Agreement between the European Communities and their Member States, of the One Part, and the Republic of Moldova, of the Other Part (98/401/EC, ECSC, Euratom)* [OJ 1998, L 181/1].

——— (1998b), *Council and Commission Decision of 26 January 1998 on the Conclusion of the Partnership and Cooperation Agreement between the European Communities and their Member States, of the One Part, and Ukraine, of the Other Part (98/149/EC, ECSC, Euratom)* [OJ 1998, L 49/1].

——— (1999), *Council and Commission Decision of 31 May 1999 on the Conclusion of the Partnership and Cooperation Agreement between the European Communities and their Member States, of the One Part, and the Republic of Armenia, of the Other Part (1999/602/EC, ECSC, Euratom)* [OJ 1999, L 239/1].

——— (1999a), *Council and Commission Decision of 31 May 1999 on the Conclusion of the Partnership and Cooperation Agreement between the European Communities and their Member States, of the One Part, and the Republic of Azerbaijan, of the Other Part (99/614/EC, ECSC, Euratom)* **[OJ 1999, L 246/1]**.

——— (1999b), *Council and Commission Decision of 31 May 1999 on the Conclusion of the Partnership and Cooperation Agreement between the European Communities and their Member States, of the One Part, and Georgia, of the Other Part (99/515/EC, ECSC, Euratom)* [OJ 1999, L 205/1].

——— (1999c), *Council and Commission Decision of 12 May 1999 on the Conclusion of the Partnership and Cooperation Agreement between the European Communities and their Member States, of the One Part, and the Republic of Kazakhstan, of the Other Part (1999/490/EC, ECSC, Euratom)* [OJ 1999, L 196/1].

——— (1999d), *Council and Commission Decision of 12 May 1999 on the Conclusion of the Partnership and Cooperation Agreement between the European Communities and their Member States, of the One Part, and the Kyrgyz Republic, of the Other Part (1999/491/EC, ECSC, Euratom)* [OJ 1999, L 196/46].

——— (1999e), *Council and Commission Decision of 31 May 1999 on the Conclusion of the Partnership and Cooperation Agreement Establishing a Partnership between the European Communities and their Member States, of the One Part, and the Republic of Uzbekistan, of the Other Part (1999/593/EC, ECSC, Euratom)* [OJ 1999, L 229/1].

——— (2009), *Decision of the Council and of the Commission of 17 November 2009 on the Conclusion of a Partnership and Cooperation Agreement Establishing a Partnership between the European Communities and their Member States, of the One Part, and the Republic of Tajikistan, of the Other Part (2009/989/EC, Euratom)* [OJ 2009, L 350/1].

Court of Justice of the European Union. (2011), *The Directive Including Aviation Activities in the EU's Emissions Trading Scheme is Valid*, Press Release (No. 139/11, December 21, 2011), Luxembourg.

——— (2012), *The Court confirms that the Commission Exceeded its Powers by Imposing on Poland and Estonia a Ceiling on Greenhouse Gas Emission Allowances*, Press Release (No. 35/12, March 29, 2012), Luxembourg.

——— (2012a), *Judgment of the Court (Grand Chamber) of 21 December 2011 (Reference for a Preliminary Ruling from the High Court of Justice, Queen's Bench Division (Administrative Court) — United Kingdom) — Air Transport Association of America, American Airlines Inc., Continental Airlines Inc., United Airlines Inc. v Secretary of State for Energy and Climate Change (Case C-366/10), Luxembourg.* See: OJ 2012, C 49/7.

——— (2013), *A Standard Term in Consumer Contracts Remains Subject to Review for Unfairness Even If It Does No More than Reproduce National Legislation Applicable to a Different Category of Contracts*, Press Release (No. 36/13, 21 March 2013), Luxembourg.

Crane, H., Kinderman, E. and Malhorta, R. (2010), *A Cubic Mile of Oil: Realities and Options for Averting the Looming Global Energy Crisis*, Oxford: Oxford University Press.

Daintith, T. and Hanch, L. (1986), *Energy Strategy in Europe: The Legal Framework*, New York: Walter de Gruyter.

de Esteban, F. (2002), 'The Future of Nuclear Energy in the European Union', Speech, Brussels, 23 May.

de Jong, S. (2012), *Europe's Southern Gas Corridor: Central Asia and the EU's Drive Towards Energy Diversification*, EU-GRASP Policy Brief, Issue 8, March 2012, http://collections.unu.edu/eserv/UNU:2907/Europes_Southern_Gas_Corridor.pdf [1 August 2015].

de Jong, J. and van der Linde, C. (2008), *EU Energy Policy in a Supply-Constrained World*, European Policy Analysis No. 11, Stockholm: Swedish Institute for European Policy Studies.

Die Welt. (2008), *Putin and Chavez Plug 'Multi-Polar World'* [26 September 2008], http://www.welt.de/english-news/article2495684/Putin-and-Chavez-plug-multi-polar-world.html [23 March 2014].

Dreyer, I., Erixon, F. and Winkler, R. (2010), *The Quest for Gas Market Competition. Fighting Europe's Dependence on Russian Gas More Effectively*, Occasional Paper, ECIPE.

ECOFYS et al. (2012), *Renewable Energy Progress and Biofuel Sustainability*, Brussels, http://ec.europa.eu/energy/renewables/reports/doc/2013_renewable_energy_progress.pdf [25 October 2013].

EIA. (2005), *International Energy Glossary*, http://www.eia.doe.gov/emeu/iea/glossary.html [30 April 2015].

——— (2012), *Effect of Increased Natural Gas Exports on Domestic Energy Markets as Requested by the Office of Fossil Energy*, Washington, DC: US Energy Information Agency.

——— (2013), *Residential Heating Oil Prices: What Consumers Should Know*, http://www.eia.gov/energyexplained/index.cfm?page=heating_oil_prices [30 March 2014].

——— (2014), *Annual Energy Outlook, Early Release*, Washington, DC: US Energy Information Agency.

Eising, R. and Jablko, N. (2001), 'Moving Targets', *Comparative Political Studies*, 34(7), 742–767.

Ellerman, A.D. and Joskow, P. (2008), *The European Union's Emissions Trading System in Perspective*, Arlington, VA: Pew Center on Global Climate Change.

Energy Charter Secretariat (ECS). (2004), *Energy Charter Treaty and Related Documents*, Brussels, http://www.encharter.org/fileadmin/user_upload/document/EN.pdf#page=211 [30 April 2014].

——— (2008), *Fostering LNG Trade*, Brussels: Energy Charter Secretariat.

Energy Community. (2013), *Decision of the Ministerial Council of the Energy Community, (DtzO13tO3lMG-EnC) on Extending the Duration of the Energy Community Treaty*, https://www.energy-community.org/pls/portal/docs/2386188.PDF [30 April 2015].

——— (2014), *Regulation on Conditions for Access to the Network for Cross-Border Exchanges in Electricity 1228/2003* [OJ 2003, L 176/1].

ENTSOG. (2013), *ENTSOG – The European Natural Gas Network (Capacities at Cross-Border Points on the Primary Market)*, Brussels: European Network of Transmission System Operators for Gas, http://www.entsog.eu/maps/transmission-capacity-map [20 February 2014].

——— (2013a), *Ten-Year Network Development Plan 2013–2022*, Brussels: European Network of Transmission System Operators for Gas.

Ernst & Young Global Limited (EYMG). (2013), *Shale Gas in Europe: Revolution or Evolution?* Earnst and Young, London, http://www.ey.com/Publication/vwLUAssets/Shale_gas_in_Europe_revolution_or_evolution/$FILE/EY-Shale_gas_in_Europe-revolution_or_evolution.pdf [20 October 2014].

EurActiv. (2005), *Consumer Electrics Switch to Energy-Efficiency Mode* [14 April 2005], http://www.euractiv.com/climate-environment/consumer-electrics-switch-energy-efficiency-mode/article-137900 [30 April 2015].

——— (2007), *Interview with Andris Piebalgs: The Commission's Energy Package* [17 January 2007], http://www.euractiv.com/en/energy/andris-piebalgs-commission-energy-package/article-160959 [30 April 2014].

——— (2008), *Deal Clinched on Capping Aircraft Emissions* [27 June 2008], http://www.euractiv.com/transport/deal-clinched-capping-aircraft-e-news-220164 [28 August 2013].

——— (2009), *Liberalising the EU Energy Sector* [7 July 2009], http://www.euractiv.com/energy/liberalising-eu-energy-sector-linksdossier-188362 [25 March 2014].

——— (2011), *Brussels Slaps Opt-Outs on Efficiency Plan*, http://www.euractiv.com/energy-efficiency/brussels-slaps-opt-outs-efficien-news-505877 [7 November 2013].

——— (2012), *Energy Efficiency Directive: Completing an Energy Policy Puzzle* [2 November 2012], http://www.euractiv.com/energy-efficiency/energy-efficiency-directive-linksdossier-514483 [3 March 2014].

——— (2013), *Biofuels Industry Sent 'Three Mails an Hour' in ILUC Lobby Offensive* [7 May 2013], http://www.euractiv.com/energy/biofuels-industry-sent-mails-hou-news-519531 [24 February 2014].

Eurelectric. (2007), *Eurelectric Views on Unbundling DSOs*, Position Paper, July 2007, http://www.eurelectric.org/media/43203/dsounbunupdmpcb160707final-2007-233-0019-2-.pdf [30 April 2014].

Eurogas. (2006), *The European Gas Market: Eurogas Views on the Way Forward*, Position Paper, 14 July 2006, http://www.europarl.europa.eu/RegData/etudes/etudes/join/2006/375866/IPOL-JOIN_ET%282006%29375866_EN.pdf [24 February 2015].

——— (2013), *Statistical Report 2013*, Brussels: Eurogas.

Europe's Energy Portal. (2014), http://www.energy.eu/fuelprices/ [6 May 2014].

European Alliance of Energy Intensive Industries. (2004), *Energy Intensive Industries' Position on the Amendments to be Voted by the EP's Environment Committee on 15 March 2003 on the Emission Trading Linking Directive*, 11 March 2004.

——— (2006), *Urgent Measures are Required to Improve the Functioning of Electricity and Gas Markets*, Position Paper, 22 September 2006.

——— (2007), *Emission Trading Scheme for NO_x and SO_2*, Position Paper, March 2007.

European Commission. (1968), *First Guidelines for a Community Energy Policy. Memorandum Presented by the Commission to the Council,*

COM (68) 1040 Final, 18 December 1968. Bulletin of the European Communities, Supplement 12/1968, http://aei.pitt.edu/5134/1/5134.pdf [30 April 2015].

——— (1972), *Necessary Progress in Community Energy Polic., Communication from the Commission to the Council, COM (72) 1200 Final, Brussels. Bulletin of the European Communities Supplement 11/1972*, http://aei.pitt.edu/5146/1/5146.pdf [30 April 2015].

——— (1973), *Guidelines and Priority Actions under the Community Energy Policy. SEC (73) 1481 Final, 19 April 1973. Bulletin of the European Communities, Supplement 6/1973*, http://aei.pitt.edu/1520/1/energy_SEC_73_1481.pdf [30 April 2014].

——— (1980a), *Economic Aspects of a Community Initiative on Energy. Communication from the Commission to the Council, COM (80) 151 Final, 20 March 1980*, http://aei.pitt.edu/32760/1/COM_(80)_151_final.pdf [30 April 2014].

——— (1980b), *Energy Price and Tax Harmonization in the Community. Communication from the Commission to the Council and the European Parliament, COM (80) 152 Final, Brussels*, http://aei.pitt.edu/5463/1/5463.pdf [30 April 2014].

——— (1980c), *Fiscal Instruments for Raising Community Revenues from Energy. Communication from the Commission to the Council and the European Parliament, COM (80) 153 Final, 20 March 1980*, http://aei.pitt.edu/5464/1/5464.pdf [30 April 2015].

——— (1980d), *A Community Programme for the Promotion of Investment in Energy. Communication from the Commission to the Council, COM (80) 220 Final, 22 April 1980*, http://aei.pitt.edu/32764/1/COM_(80)_220_final.pdf [30 April 2014].

——— (1980e), *Energy Policy. Communication from the Commission to the European Council, 31st March–1st April 1980, COM (80) 130 Final, 20 March 1980*, http://aei.pitt.edu/31881/1/COM_(80)_130_final.pdf [30 April 2014].

——— (1980f), *Energy and Economic Policy, COM (80) 583 Final, 15 October 1980*, http://aei.pitt.edu/34131/1/COM_(80)_583_final.pdf [30 April 2014].

——— (1981a), *Review of Energy Policy Objectives for 1990 and Member States' Investment Programmes. Communication from the Commission to the Council, COM (81) 64 Final, 24 February 1981. Bulletin of the European Communities*, 2/1981, 1.5.1–1.5.6, http://aei.pitt.edu/31884/1/COM_(81)_64_final.pdf [30 April 2014].

——— (1981b), *The Development of an Energy Strategy for the Community. Communication from the Commission to the Council, Mandate of 30 May 1980, COM (81) 540 Final, 2 October 1981. Bulletin of the European Communities, Supplement 4/1981*, pp. 7–20, http://aei.pitt.edu/1508/1/energy_30_May_COM_81_540.pdf [30 April 2014].

——— (1982), *Proposal for a Council Regulation on the Payment of Financial Incentives in Support of Categories of Investment in the Rational Use of Energy, COM (82) 357 Final* [OJ 1982, C 285/5].

——— (1983), *Energy and Energy Research in the Community: A Five-Year Programme of Action and its Financing. Communication from the Commission, COM (83) 315 Final, 15 June 1983. Bulletin of the European Communities, Supplement 5/1983*, pp. 12–24, http://aei.pitt.edu/5880/1/5880.pdf [30 April 2014].

——— (1983a), *Energy Strategy: A Five-Year Programme, Bulletin of the European Communities, 6/1983, Brussels*, pp. 11–14 [OJ 1983, C 68].

——— (1985), *Completing the Internal Market, COM (85) 310 Final, Brussels,* http://europa.eu/documents/comm/white_papers/pdf/com1985_0310_f_en.pdf [8 December 2014].

——— (1988), *The Internal Energy Market. Commission Working Document, COM (88) 238 Final, 2 May 1988*, http://aei.pitt.edu/4037/1/4037.pdf [30 April 2014].

——— (1990), *Communication from the Commission to the Council on Energy and the Environment, COM (89) 369 Final, 8 February 1990*, http://eur-lex.europa.eu/legal-content/EN/TXT/PDF/?uri=CELEX:51989DC0369&qid=1432473275607&from=EN [30 April 2015].

——— (1992), *Proposal for a Council Directive Introducing a Tax on Carbon Dioxide Emissions and Energy, COM (92) 226 Final, Brussels* [OJ 1992, C 196/1].

——— (1995), *For a European Union Energy Policy – Green Paper, COM (94) 659 Final/2, 23 February 1995*, http://aei.pitt.edu/1185/1/energy_gp_COM_94_659.pdf [30 April 2014].

——— (1995a), *White Paper: An Energy Policy for the European Union, COM (95) 682 Final, 13 December 1995*, http://eur-lex.europa.eu/legal-content/EN/TXT/PDF/?uri=CELEX:51995DC0682&qid=1432474072400&from=EN [30 April 2014].

——— (1995b), *A Community Strategy to Reduce CO_2 Emissions from Passenger Cars and Improve Fuel Economy*, Communication, *COM (95) 689 Final*, Brussels.

——— (1996), *Energy for the Future: Renewable Sources of Energy, Green Paper for a Community Strategy, COM (96) 576 Final, 20 November 1996*, http://eur-lex.europa.eu/legal-content/EN/TXT/PDF/?uri=CELEX:51996DC0576&rid=1 [30 April 2014].

——— (1997), *Energy for the Future: Renewable Sources of Energy. White Paper for a Community Strategy and Action Plan, COM (97) 599 Final, 26 November 1997*, http://europa.eu/documents/comm/white_papers/pdf/com97_599_en.pdf [30 April 2014].

——— (1998), *Strengthening Environmental Integration within Community Energy Policy, COM (98) 571 Final, 14 October 1998*, http://aei.pitt.edu/3371/1/3371.pdf [30 April 2014].

——— (1999), *Preparing for Implementation of the Kyoto Protocol. Communication from the Commission to the Council and the European Parliament, COM (99) 230 Final, 19 June 1999*, http://aei.pitt.edu/4951/1/4951.pdf [30 April 2014].

——— (2000), *Towards a European Strategy for Energy Supply Security – Green Paper, COM (2000) 769 Final, 29 November 2000*, http://eur-lex.europa.eu/legal-content/EN/TXT/HTML/?uri=CELEX:52000DC0769&rid=1 [30 April 2015].

——— (2000a), *Green Paper on Greenhouse Gas Emissions Trading Within the European Union, COM (2000) 87 Final, 8 March 2000*, http://eur-lex.europa.eu/legal-content/EN/TXT/PDF/?uri=CELEX:52000DC0087&rid=1 [30 April 2014].

——— (2000b), *Communication from the Commission to the Council, The European Parliament, The Economic and Social Committee and the Committee of the Regions: Action Plan to Improve Energy Efficiency in the European Community, COM (2000) 247 Final, 26 April 2000*, http://eur-lex.europa.eu/legal-content/EN/TXT/PDF/?uri=CELEX:52000DC0247&rid=1 [30 April 2014].

——— (2001), *European Transport Policy for 2010: Time to Decide – White Paper,. COM (2001) 370 Final, 12 September 2001*, http://aei.pitt.edu/1187/1/transport_2010_wp_com_2001_370.pdf [30 April 2014].

——— (2001a), *Proposal for a Directive Establishing a Scheme for Greenhouse Gas Emission Allowance Trading within the Community and Amending Council Directive 96/91/EC, COM (2001) 581 Final, 23 October 2001* [OJ 2002, C 75 E/33].

——— (2001b), *European Governance*, White Paper, COM *(2001) 428 Final, Brussels* [OJ 2001, C 287/1].

——— (2001c), *Commission Decision of 11 July 2001 Setting Up a Consultative Committee, to be known as the 'European Energy and Transport Forum', 2001/546/EC* [OJ 2001, L 195/58].

——— (2002) *Nuclear Safety in the European Union, Communication COM (2002) 605 Final, 6 November 2002*, http://eur-lex.europa.eu/legal-content/EN/TXT/PDF/?uri=CELEX:52002DC0605&rid=1 [30 April 2014].

——— (2002a), *Communication from the Commission of 26 June 2002, Final Report on the Green Paper towards a European Strategy for the Security of Energy Supply, COM(2002) 321 Final*, http://eur-lex.europa.eu/legal-content/EN/TXT/PDF/?uri=CELEX:52002DC0321&rid=1 [30 April 2014].

——— (2002b), *Communication from the Commission of 21 February 2002, Towards a Global Partnership for Sustainable Development, COM (2002) 82 Final*, http://eur-lex.europa.eu/legal-content/EN/TXT/PDF/?uri=CELEX:52002DC0082&rid=2 [30 April 2014].

——— (2003), *Decision of 11 November 2003 on Establishing the European Regulators Group for Electricity and Gas, 2003/796/EC* [OJ 2003, L 296/34].

——— (2003a), *Proposal for a Directive of the European Parliament and of the Council on Establishing a Framework for the Setting of Eco-Design Requirements for Energy-Using Products and Amending Council Directive 92/42/EEC, COM (2003) 453 Final, 1 August 2003*, http://eur-lex.europa.eu/legal-content/EN/TXT/PDF/?uri=CELEX:52003PC0453&rid=1 [30 April 2014].

——— (2004), *Repertoire of the Acquis Communautaire. Energy and Transport*, 31 December 2004, http://ec.europa.eu/dgs/energy_transport/acquis/acquis_energy_transport_en.pdf [30 April 2015].

——— (2004a), *Commission Recommendation of 18 December 2003 on Standardised Information on Radioactive Airborne and Liquid Discharges into the Environment from Nuclear Power Reactors and Reprocessing Plants in Normal Operation, 2004/2/Euratom* [OJ 2004, L 2/36].

——— (2005), *Green Paper on Energy Efficiency or Doing More with Less, COM (2005) 265, 22 June 2005*, http://eur-lex.europa.eu/legal-content/EN/TXT/PDF/?uri=CELEX:52005DC0265&rid=1 [30 April 2014].

——— (2005a), *Reducing the Climate Change Impact of Aviation, Report on the Public Consultation March–May 2005*, http://ec.europa.eu/clima/policies/transport/aviation/docs/report_publ_cons_en.pdf [30 April 2014].

——— (2006), *A European Strategy for Sustainable, Competitive and Secure Energy, Green Paper, COM (2006) 105 Final, 8 March 2006*, http://eur-lex.europa.eu/legal-content/EN/TXT/PDF/?uri=CELEX:52006DC0105&rid=1 [30 April 2014].

——— (2006a), *Inquiry Pursuant to Article 17 of Regulation (EC) No 1/2003 into the European Gas and Electricity Sectors (Final Report), Communication from the Commission, COM (2006) 851 Final*, Brussels.

——— (2006b), *Action Plan for Energy Efficiency: Realising the Potential, Communication from the Commission, COM (2006) 545 Final, 19 October 2006*, http://eur-lex.europa.eu/legal-content/EN/TXT/PDF/?uri=CELEX:52006DC0545&rid=1 [30 April 2014].

——— (2006e), *Proposal for a Directive of the European Parliament and of the Council Amending Directive 2003/87/EC so as to Include Aviation Activities in the Scheme for Greenhouse Gas Emission Allowance Trading within the Community, COM (2006) 818 Final, 20 December 2006*, http://eur-lex.europa.eu/legal-content/EN/TXT/PDF/?uri=CELEX:52006PC0818&rid=1 [30 April 2014].

——— (2006f), *Ministerial Declaration on Enhanced Energy Co-operation between the EU, the Littoral States of the Black and Caspian Seas and their Neighbouring Countries*, http://ec.europa.eu/dgs/energy_transport/international/regional/caspian/doc/2006_11_30_astana_conclusions.pdf [30 April 2015].

——— (2007), *An Energy Policy for Europe, Communication from the Commission to the European Council and the European Parliament, COM (2007) 1 Final, 10 January 2007*, http://eur-lex.europa.eu/legal-content/EN/TXT/PDF/?uri=CELEX:52007DC0001&rid=1 [30 April 2006].

——— (2007a), *A European Strategic Energy Technology Plan (SET Plan) – Towards a Low Carbon Future, Communication from the Commission, COM (2007) 723 Final, 22 November 2007*, http://eur-lex.europa.eu/legal-content/EN/TXT/PDF/?uri=CELEX:52007DC0723&rid=1 [30 April 2006].

——— (2007b) *A European Approach to Nuclear Power, Safety and Security European Commission – MEMO/07/10, 10 January 2007*, http://europa.eu/rapid/press-release_MEMO-07-10_en.pdf [30 April 2014].

——— (2007c), *Prospects for the Internal Gas and Electricity Market, Communication, COM (2006) 841 Final, 10 January 2007*, http://eur-lex.europa.eu/legal-content/EN/TXT/PDF/?uri=CELEX:52006DC0841&from=EN [30 April 2014].

——— (2007d), *Commission Energy Sector Inquiry Confirms Serious Competition Problems*, Press Release of 10 January 2007, IP/07/26, http://europa.eu/rapid/press-release_IP-07-26_en.pdf [30 April 2014].

——— (2007e), *Proposal for a Directive Amending Directive 2003/54/EC Concerning Common Rules for the Internal Market in Electricity, COM (2007) 528 Final, 19 September 2007*, http://eur-lex.europa.eu/legal-content/EN/TXT/PDF/?uri=CELEX:52007PC0528&rid=1 [30 April 2014].

——— (2007f), *Proposal for a Directive Amending Directive 2003/55/EC Concerning Common Rules for the Internal Market in Natural Gas, COM (2007) 529, 19 September 2007*, http://eur-lex.europa.eu/legal-content/EN/TXT/PDF/?uri=CELEX:52007PC0529&rid=1 [30 April 2014].

——— (2007g), *Proposal for a Regulation Amending Regulation (EC) No. 1775/2005 Establishing an Agency for the Cooperation of Energy Regulators, COM (2007) 530, 19 September 2007*, http://eur-lex.europa.eu/legal-content/EN/TXT/PDF/?uri=CELEX:52007PC0530&rid=1 [30 April 2014].

——— (2007h), *Proposal for a Regulation Amending Regulation (EC) No. 1228/2003 on Conditions for Access to the Network for Cross-Border Exchanges in Electricity, COM (2007) 531, 19 September 2007*, http://eur-lex.europa.eu/legal-content/EN/TXT/PDF/?uri=CELEX:52007PC0531&rid=1 [30 April 2014].

——— (2007i), *Proposal for a Regulation Amending Regulation (EC) No. 1775/2005 on Conditions for Access to the Natural Gas Transmission*

Networks, COM (2007) 532, 19 September 2007, http://eur-lex.europa.eu/legal-content/EN/TXT/PDF/?uri=CELEX:52007PC0532&rid=1 [30 April 2014].

——— (2007j), *DG Competition Report of 10 January 2007 on Energy Sector Inquiry, SEC (2006) 1724, 10 January 2007*, http://ec.europa.eu/competition/sectors/energy/2005_inquiry/full_report_part1.pdf [30 April 2014].

——— (2007k), *Proposal for a Regulation of The European Parliament and of The Council Setting Emission Performance Standards for New Passenger Cars as Part of the Community's Integrated Approach to Reduce CO_2 Emissions from Light-Duty Vehicles, COM (2007) 856 Final, 19 December 2007*, http://eur-lex.europa.eu/legal-content/EN/TXT/PDF/?uri=CELEX:52007PC0856&rid=1 [30 April 2014].

——— (2007l), *Sustainable Power Generation from Fossil Fuels: Aiming for Near-Zero Emissions from Coal after 2020, Communication from the Commission, COM (2006) 843 Final, 10 January 2007*, http://eur-lex.europa.eu/legal-content/EN/TXT/PDF/?uri=CELEX:52006DC0843&rid=1 [30 April 2015].

——— (2007m), *Renewable Energy Road Map, Renewable Energies in the 21st Century: Building a More Sustainable Future, Communication from the Commission, COM (2006) 848 Final, 10 January 2007*, http://eur-lex.europa.eu/legal-content/EN/TXT/PDF/?uri=CELEX:52006DC0848&rid=1 [30 April 2015].

——— (2008), *Questions and Answers on the Commission's Proposal to Revise the EU Emissions Trading System, MEMO/08/35, 23 January 2008*, http://europa.eu/rapid/press-release_MEMO-08-35_en.htm [30 April 2014].

——— (2008a), *20 20 by 2020 – Europe's Climate Change Opportunity, Communication from The Commission to The European Parliament, The Council, The European Economic and Social Committee and The Committee of The Regions, COM (2008) 30 Final, 23 January 2008*, http://eur-lex.europa.eu/legal-content/EN/TXT/PDF/?uri=CELEX:52008DC0030&rid=1 [30 April 2014].

——— (2008b), *Proposal for Directive on the Promotion of the Use of Energy from Renewable Sources, COM (2008) 19 Final, 23 January 2008*, http://eur-lex.europa.eu/legal-content/EN/TXT/PDF/?uri=CELEX:52008PC0019&rid=1 [30 April 2014].

——— (2008c), *Second Strategic Energy Review — An EU Energy Security and Solidarity Action Plan, COM (2008) 781 Final, 13 November 2008*, http://eur-lex.europa.eu/legal-content/EN/TXT/PDF/?uri=CELEX:52008DC0781&rid=2 [30 April 2014].

——— (2009), *Antitrust: Commission Fines E.ON and GDF Suez €553 Million Each for Market-Sharing in French and German Gas Markets*, Press Release of 8 July 2009, IP/09/1099, http://europa.eu/rapid/press-release_IP-09-1099_en.htm [30 April 2014].

——— (2009a), *Report on Progress in Creating the Internal Gas and Electricity Market, COM (2009) 115 Final, 11 March 2009*, http://eur-lex.europa.eu/legal-content/EN/TXT/PDF/?uri=CELEX:52009DC0115&rid=1 [30 April 2014].

——— (2009b), *Commission Regulation of 18 March 2009 Implementing Directive 2005/32/EC of the European Parliament and of the Council with Regard to Ecodesign Requirements for Non-directional Household Lamps, (EC)* No. *244/2009, Brussels* [OJ 2009, L 76/3].

——— (2009c), *Commission Regulation of 18 March 2009 Implementing Directive 2005/32/EC of the European Parliament and of the Council with Regard to Ecodesign Requirements for Fluorescent Lamps without Integrated Ballast, for High Intensity Discharge Lamps, and for Ballasts and Luminaires Able to Operate Such Lamps, and Repealing Directive 2000/55/EC of the European Parliament and of the Council, (EC)* No. *245/2009, Brussels* [OJ 2009, L 76/17].

——— (2009d), *Commission Recommendation of 11 February 2009 on the Implementation of a Nuclear Material Accountancy and Control System by Operators of Nuclear Installations* [OJ 2009, L 41/17].

——— (2009e), *The January 2009 Gas Supply Disruption to The EU: An Assessment, Commission Staff Working Document. Accompanying document to the 'Proposal for a Regulation of The European Parliament and of The Council Concerning Measures to Safeguard Security of Gas Supply and Repealing Directive 2004/67/EC', SEC(2009) 977, 16 July 2009*, http://eur-lex.europa.eu/legal-content/EN/TXT/PDF/?uri=CELEX:52009SC0977&rid=1 [30 April 2009].

——— (2009f), *Press Release, Union for the Mediterranean: Commission Increases its Contribution to Priority Projects,* 10 July 2009, IP/09/1113, http://europa.eu/rapid/press-release_IP-09-1113_en.htm?locale=en [30 April 2014].

——— (2009g), *Memorandum on an Early Warning Mechanism in the Energy Sector within the Framework of the EU-Russia Energy Dialogue, 16 November 2009*, http://ec.europa.eu/energy/sites/ener/files/documents/2009_11_16_ewm_signed_en_0.pdf [30 April 2015].

——— (2010), *Commission Regulation (EU)* 920/2010 *of 7 October 2010 for a Standardised and Secured System of Registries Pursuant to Directive 2003/87/EC of the European Parliament and of the Council and Decision No. 280/2004/EC of the European Parliament and of the Council* [OJ 2010, L 270/1].

——— (2010a), *Energy 2020 – A Strategy for Competitive, Sustainable and Secure Energy*, COM *(2010) 639 Final, 10 November 2010*, http://eur-lex.europa.eu/legal-content/EN/TXT/PDF/?uri=CELEX:52010DC0639&rid=3 [30 April 2014].

——— (2010b), *Energy Infrastructure Priorities for 2020 and Beyond – A Blueprint for an Integrated European Energy Network*, COM *(2010)*

677 Final, 17 November 2010, http://eur-lex.europa.eu/legal-content/EN/TXT/PDF/?uri=CELEX:52010DC0677&from=EN [30 April 2014].

——— (2010c), *Special Eurobarometer 324, Europeans and Nuclear Safety, March 2010*, http://ec.europa.eu/public_opinion/archives/ebs/ebs_324_en.pdf [30 April 2014].

——— (2011), *Commission Regulation (EU) No. 1210/2011 of 23 November 2011 Amending Regulation (EU) No. 1031/2010 in Particular to Determine the Volume of Greenhouse Gas Emission Allowances to be Auctioned Prior to 2013* [OJ 2011, L 308/2].

——— (2011a), *A Roadmap for Moving to a Competitive Low Carbon Economy in 2050, COM (2011) 112 Final, 8 March 2011*, http://eur-lex.europa.eu/legal-content/EN/TXT/PDF/?uri=CELEX:52011DC0112&rid=2 [30 April 2014].

——— (2011b), *Energy Roadmap 2050, COM (2011) 885 Final, 15 December 2011*, http://eur-lex.europa.eu/legal-content/EN/TXT/PDF/?uri=CELEX:52011DC0885&rid=1 [30 April 2014].

——— (2011c), *Roadmap to a Single European Transport Area – Towards a Competitive and Resource Efficient Transport System, White Paper, COM (2011) 144 Final, 28 March 2011*, http://eur-lex.europa.eu/legal-content/EN/TXT/PDF/?uri=CELEX:52011DC0144&rid=3 [30 April 2014].

——— (2011d), *Energy Efficiency Plan 2011, Communication, COM (2011) 109, 8 March 2011*, http://eur-lex.europa.eu/legal-content/EN/TXT/PDF/?uri=CELEX:52011DC0109&rid=2 [30 April 2014].

——— (2011e), *Proposal for a Directive on Energy Efficiency and Repealing Directives 2004/8/EC and 2006/32/EC, COM (2011) 370 Final, 22 June 2011*, http://eur-lex.europa.eu/legal-content/EN/TXT/PDF/?uri=CELEX:52011PC0370&rid=1 [30 April 2014].

——— (2011f), *Proposal for a Council Directive Amending Directive 2003/96/EC Restructuring the Community Framework for the Taxation of Energy Products and Electricity, COM (2011) 169/3, 13 April 2011*, http://eur-lex.europa.eu/legal-content/EN/TXT/PDF/?uri=CELEX:52011PC0169&rid=1 [30 April 2014].

——— (2011g), *The EU Energy Policy: Engaging with Partners beyond Our Borders, COM (2011) 539, 7 September 2011*, http://eur-lex.europa.eu/legal-content/EN/TXT/PDF/?uri=CELEX:52011DC0539&rid=1 [30 April 2014].

——— (2011h), *Public Consultation Document: The External Dimension of the EU Energy Policy*, http://ec.europa.eu/energy/international/consultations/doc/20110221_consultation_document.pdf [20 February 2014].

——— (2011i), *Energy Infrastructure. Priorities for 2020 and Beyond – A Blueprint for an Integrated European Energy Network*, https://ec.europa.eu/energy/sites/ener/files/documents/2011_energy_infrastructure_en.pdf [30 April 2014].

——— (2011j), *Proposal for a Decision Setting Up an Information Exchange Mechanism with Regard to Intergovernmental Agreements between Member States and Third Countries in the Field of Energy, COM (2011) 540, 7 September 2011*, http://eur-lex.europa.eu/legal-content/EN/TXT/PDF/?uri=CELEX:52011PC0540&rid=1 [30 April 2014].

——— (2011k), *Press Release, Energy Taxation: Commission Promotes Energy Efficiency and More Environmental Friendly Products*, 13 April 2011, IP/11/468, http://europa.eu/rapid/press-release_IP-11-468_en.htm [30 April 2014].

——— (2011l), *Memorandum on a Mechanism for Preventing and Overcoming Emergency Situations in the Energy Sector within the Framework of the EU-Russia Energy Dialogue (Early Warning Mechanism)*, http://ec.europa.eu/energy/sites/ener/files/documents/20110224_memorandum.pdf [30 April 2015].

——— (2012), *Statistical Pocketbook 2012: Transport in Figures*, http://ec.europa.eu/transport/facts-fundings/statistics/doc/2012/pocketbook2012.pdf [1 September 2013].

——— (2012a), *Proposal for a Decision of the European Parliament and of the Council Derogating Temporarily from Directive 2003/87/EC of the European Parliament and of the Council Establishing a Scheme for Greenhouse Gas Emission Allowance Trading within the Community, COM (2012) 697 Final, 20 November 2012*, http://eur-lex.europa.eu/legal-content/EN/TXT/PDF/?uri=CELEX:52012PC0697&rid=1 [30 April 2014].

——— (2012b), *Proposal for a Directive Amending Directive 98/70/EC Relating to the Quality of Petrol and Diesel Fuels and Amending Directive 2009/28/EC on the Promotion of the Use of Energy from Renewable Sources, COM (2012) 595 Final, 17 October 2012*, http://eur-lex.europa.eu/legal-content/EN/TXT/PDF/?uri=CELEX:52012PC0595&rid=2 [30 April 2014].

——— (2012c), *Joint Communication to The European Parliament and The Council, Developing a European Union Policy towards the Arctic Region: Progress Since 2008 and Next Steps, JOIN (2012) 19 Final, 26 June 2012*, http://eur-lex.europa.eu/legal-content/EN/TXT/PDF/?uri=CELEX:52012JC0019&rid=1 [30 April 2014].

——— (2012d), *Investment Projects in Energy Infrastructure, Accompanying the Communication 'Making the Internal Energy Market Work' (COM (2012) 663), SWD (2012) 367, 15 November 2012*, http://eur-lex.europa.eu/legal-content/EN/TXT/PDF/?uri=CELEX:52012DC0663&rid=1 [30 April 2014].

——— (2012e), *Commission Staff Working Document: Proportionate Impact Assessment*, http://ec.europa.eu/clima/policies/ets/cap/auctioning/docs/swd_2012_xx2_en.pdf [15 September 2014].

——— (2012f), *Euratom FP7 Research & Training Projects, Volume 3, EUR 24859 EN*, http://ec.europa.eu/research/energy/euratom/publications/pdf/

euratom_fp7_research_&_training_projects_volume_3.pdf [30 April 2014].
——— (2012g), *Renewable Energy and Energy Efficiency Project Preparation Initiative in Support of Mediterranean Solar Plan (MSP-PPI)*, http://ec.europa.eu/europeaid/blending/renewable-energy-and-energy-efficiency-project-preparation-initiative-support-mediterranean_en [30 April 2015].
——— (2012h), *Press Release: Commissioner Oettinger Welcomes TANAP Gas Pipeline Agreements, Brussels*, 27 June 2012, IP/712/721, http://europa.eu/rapid/press-release_IP-12-721_en.htm [30 April 2015].
——— (2013), *Statistical Pocketbook 2013: Transport in Figures*, http://ec.europa.eu/transport/facts-fundings/statistics/pocketbook-2013_en.htm, [20 May 2014].
——— (2013a), *Commission Regulation (EU) No. 397/2013 of 30 April 2013 Amending Regulation (EC) No. 443/2009 of the European Parliament and of the Council as Regards the Monitoring of CO_2 Emissions from New Passenger Cars* [OJ 2013, L 120/4].
——— (2013b), *EU Energy in Figures, Statistical Pocketbook 2013*, Brussels: European Union.
——— (2013c), *Quarterly Report on European Gas Markets, Market Observatory for Energy, DG Energy, Volume 6(2), Second Quarter 2013*, http://ec.europa.eu/energy/sites/ener/files/documents/20130814_q2_quarterly_report_on_european_gas_markets.pdf [30 April 2014].
——— (2013d), *Renewable Energy Progress Report, COM (2013) 175 Final, 27 March 2013*, http://eur-lex.europa.eu/legal-content/EN/TXT/PDF/?uri=CELEX:52013DC0175&rid=1 [30 April 2014].
——— (2013e), *Commission Decision of 05.09.2013 on the Conclusion of a Memorandum of Understanding for a Partnership between the European Atomic Energy Community and the International Atomic Energy Agency on Nuclear Safety Cooperation, COM (2013) 5641 Final, 5 September 2013*, https://ec.europa.eu/energy/sites/ener/files/documents/2013_5641_f1_commission_decision_en_v4_p1_738743_en.pdf [30 April 2014].
——— (2013f), *A 2030 Framework for Climate and Energy Policies, Green Paper, COM (2013) 169 Final, 27 March 2013*, http://eur-lex.europa.eu/legal-content/EN/TXT/PDF/?uri=CELEX:52013DC0169&rid=1 [30 April 2015].
——— (2013g), *Report from the Commission to the European Parliament, the Council and the European Economic and Social Committee, Implementation of the Communication on Security of Energy Supply and International Cooperation and of the Energy Council Conclusions of November 2011*, COM (2013) 0638 Final, Brussels.
——— (2013h), *Communication from the Commission to the Council and the European Parliament on the Comprehensive Risk and Safety Assessments ('stress tests') of Nuclear Power Plants in the European Union and Related Activities, COM (2012) 0571 Final*, http://eur-lex.europa.

eu/legal-content/EN/TXT/PDF/?uri=CELEX:52012DC0571&from=EN [30 April 2015].

——— (2013i), *Proposal for a Council Directive Amending Directive 2009/71/EURATOM Establishing a Community Framework for the Nuclear Safety of Nuclear Installations, COM (2013) 715 Final, 17 October 2013*, http://eur-lex.europa.eu/legal-content/EN/TXT/PDF/?uri=CELEX:52013PC0715&rid=1 [30 April 2015].

——— (2013j), *Delivering the Internal Electricity Market and Making the Most of Public Intervention, Communication from the Commission, COM (2013) 7243, 11 November 2013*, http://ec.europa.eu/energy/sites/ener/files/documents/com_2013_public_intervention_en.pdf [30 April 2014].

——— (2013k), *Paper of the Services of DG Competition Containing Draft Guidelines on Environmental and Energy Aid for 2014–2020*, http://ec.europa.eu/competition/consultations/2013_regional_aid_guidelines/paper_en.pdf [30 April 2014].

——— (2013l), *DG ENER Working Paper: The Future Role and Challenges of Energy Storage*, http://ec.europa.eu/energy/infrastructure/doc/energy-storage/2013/energy_storage.pdf [30 April 2014]

——— (2013m), *Commission Regulation (EU) 389/2013 of 2 May 2013 Establishing a Union Registry Pursuant to Directive 2003/87/EC of the European Parliament and of the Council, Decisions No. 280/2004/EC and No. 406/2009/EC of the European Parliament and of the Council and Repealing Commission Regulations (EU) No. 920/2010 and No. 1193/2011* [OJ 2013, L 122/1].

——— (2013n), *The EU Emissions Trading System (EU ETS)*, Brussels.

——— (2013o), *Consultation on Structural Options to Strengthen the EU Emissions Trading System, Multiple Documents*, http://ec.europa.eu/clima/consultations/articles/0017_en.htm [15 September 2014].

——— (2013p), *Quarterly Report on European Electricity Markets, Market Observatory for Energy, DG Energy, Volume 6(2), Second Quarter 2013*, http://ec.europa.eu/energy/sites/ener/files/documents/20130814_q2_quarterly_report_on_european_electricity_markets.pdf [30 April 2015].

——— (2013q), *Press Release: Energy: Commission Unveils List of 250 Infrastructure Projects that may Qualify for €5.85 Billion of Funding, Brussels, 14 October 2013*, http://europa.eu/rapid/press-release_IP-13-932_en.htm [25 October 2013].

——— (2014), *A Policy Framework for Climate and Energy in the Period from 2020 to 2030, Communication from The Commission to the European Parliament, the Council, the European Economic and Social Committee and the Committee of The Regions, COM (2014) 15 Final, 22 January 2014*, http://eur-lex.europa.eu/legal-content/EN/TXT/PDF/?uri=CELEX:52014DC0015&rid=3 [30 April 2015].

——— (2014a), *Communication from The Commission to the Council and the European Parliament on the Exploration and Production of*

Hydrocarbons (such as Shale Gas) Using High Volume Hydraulic Fracturing in the EU, COM (2014) 023 Final, 22 January 2014, http://eur-lex.europa.eu/legal-content/EN/TXT/PDF/?uri=CELEX:52014DC0023&rid=2 [30 April 2015].

——— (2014b), *Recommendation on Minimum Principles for the Exploration and Production of Hydrocarbons (such as Shale Gas) Using High-Volume Hydraulic Fracturing, (2014/70/EU)* [OJ 2014, L 39/72].

——— (2014c), *European Commission – Speech of Maroš Šefčovič – Vice-president for Energy Union, 'Building a Euro-Mediterranean Bridge: The Strategic Importance of Gas and Electricity Networks in the Context of Energy Security', Building a Euro Mediterranean Energy Bridge, High Level Conference, Rome*, SPEECH/14/1902, http://europa.eu/rapid/press-release_SPEECH-14-1902_en.htm [30 April 2015].

——— (2014d), *The European Files*, March–April 2014, http://europeanfiles.eu/wp-content/uploads/issues/2014march-april.pdf [30 April 2015].

——— (2014e), *Progress Towards the Internal Energy Market 2014: Pilot Project for EU Wide Electricity Trade Starts Today*, Daily News of 4 February 2014 (MEX/14/0204), http://europa.eu/rapid/midday-express-04-02-2014.htm [30 April 2015].

——— (2014f), *Commission Staff Working Document, Energy Prices and Costs Report. Accompanying the Document, Communication from the Commission to the European Parliament, the Council, and the European Economic and Social Committee and the Committee of the Regions, 'Energy Prices and Costs in Europe', SWD (2014) 20 Final, 22 January 2014*, http://eur-lex.europa.eu/legal-content/EN/TXT/PDF/?uri=CELEX:52014SC0020R(01)&rid=1 [30 April 2015].

——— (2014g), *Proposal for a Decision of the European Parliament and of the Council Concerning the Establishment and Operation of a Market Stability Reserve for the Union Greenhouse Gas Emission Trading Scheme and Amending Directive 2003/87/EC, COM (2014) 20/2 Final, 22 January 2014*, http://eur-lex.europa.eu/legal-content/EN/TXT/PDF/?uri=CELEX:52014PC0020&rid=1 [30, April 2015].

——— (2014h), *Commission Regulation (EU) No. 176/2014 of 25 February 2014 Amending Regulation (EU) No. 1031/2010 in Particular to Determine the Volumes of Greenhouse Gas Emission Allowances to be Auctioned in 2013–20* [OJ 2014, L 56/11].

——— (2014i), *Communication from The Commission: Guidelines on State Aid for Environmental Protection and Energy 2014–2020 (2014/C 200/01)* [OJ 2014, C 1-55].

——— (2014j), *EU Energy in Figures, Statistical Pocketbook 2014*, Brussels: European Union.

——— (2014k), *Press Release: Common Rules and Infrastructure Needed to Complete Energy Market*, 13 October 2014, IP/14/1124, http://europa.eu/rapid/press-release_IP-14-1124_en.htm [25 October 2015].

——— (2014l), *Commission Staff Working Document: In-Depth Study of European Energy Security. Accompanying the Document, Communication from the Commission to the Council and the European Parliament, 'European Energy Security Strategy', COM (2014) 330 Final, 28 May 2014*, http://ec.europa.eu/energy/doc/20140528_energy_security_study.pdf [20 October 2015].

——— (2014m), *Online Summary of Legislation: Green Paper on the Security of Energy Supply*, http://europa.eu/legislation_summaries/energy/external_dimension_enlargement/l27037_en.htm [30 April 2015].

——— (2014n), *European Commission Newsletter February 2014, Energy in Europe*, https://ec.europa.eu/energy/sites/ener/files/documents/Newsletter_February2014.pdf [30 April 2015].

——— (2014o), *EU Action on Climate*, http://ec.europa.eu/clima/policies/brief/eu/index_en.htm [30 April 2015]; see also (2015b).

——— (2014p), *Research & Innovation, Energy, Wind Energy*, http://ec.europa.eu/research/energy/eu/index_en.cfm?pg=research-wind [30 April 2015].

——— (2014q), *Communication from the Commission to the European Parliament and the Council on the Short Term Resilience of the European Gas System, COM (2014) 654 Final*, http://eur-lex.europa.eu/legal-content/EN/TXT/PDF/?uri=CELEX:52014DC0330&from=EN [24 February 2015].

——— (2014r), C*ommission Recommendation of 29 October 2014 on the Application of Internal Energy Market Rules between the EU Member States and the Energy Community Contracting Parties (2014/761/EU)*, http://eur-lex.europa.eu/search.html?qid=1432992142368&text=2014/761/EU&scope=EURLEX&type=quick&lang=en [23 February 2014].

——— (2015), *Phase-Out of Inefficient Lamps Postponed to 1 September 2018* [17 April 2015], https://ec.europa.eu/energy/en/news/phase-out-inefficient-lamps-postponed-1-september-2018 [30 April 2015].

——— (2015a), *Roadmap for Moving to a Low-Carbon Economy in 2050*, http://ec.europa.eu/clima/policies/roadmap/index_en.htm [30 April 2015].

——— (2015b), *EU Action on Climate*, http://ec.europa.eu/clima/policies/brief/eu/index_en.htm [30 April 2015].

——— (2015c), *The Strategic Energy Technologies Information System, the SET-PLAN Roadmap on Low Carbon Energy Technologies*, https://setis.ec.europa.eu/implementation/technology-roadmap/the-set-plan-roadmap-on-low-carbon-energy-technologies [30 April 2015].

——— (2015d), *The Strategic Energy Technologies Information System, European Industrial Initiative on Wind Energy*, https://setis.ec.europa.eu/implementation/technology-roadmap/european-industrial-initiative-on-wind-energy-1 [30 April 2015].

——— (2015e), *The Strategic Energy Technologies Information System, European Industrial Initiative on Bioenergy*, https://setis.ec.europa.eu/implementation/technology-roadmap/european-industrial-initiative-on-bioenergy [30 April 2015].

European Commission and Russian Federation. (2013), *Roadmap, EU-Russia Energy Cooperation until 2050*, http://ec.europa.eu/energy/sites/ener/files/documents/2013_03_eu_russia_roadmap_2050_signed.pdf [30 April 2014].

European Convention. (2003), *Draft Treaty Establishing a Constitution for Europe Adopted by Consensus by the European Convention on 13 June and 10 July 2003 Submitted to The President of The European Council in Rome* [OJ 2003, C 169/01].

European Council. (2003), *Council Directive of 27 October 2003 Restructuring the Community Framework for the Taxation of Energy Products and Electricity, 2003/96/EC* [OJ 2003, L 283/51].

—— (2005), *Declaration by the Heads of State or Government of the Member States of the European Union on the Ratification of the Treaty Establishing a Constitution for Europe, SN 117/05, 18 June 2005*, https://www.consilium.europa.eu/uedocs/cms_data/docs/pressdata/en/ec/85325.pdf [30 April 2014].

——— (2011), *Conclusions of 4 February 2011, ST 2 2011 REV 1*, http://data.consilium.europa.eu/doc/document/ST-2-2011-REV-1/en/pdf [10 October 2014].

——— (2013), *European Council (22 May 2013) – Draft Conclusions, ST 7625 2013 INIT, 22 May 2013*, http://data.consilium.europa.eu/doc/document/ST-7625-2013-INIT/en/pdf [30 April 2014]. See also: Commission Press Release D/13/4, 22 May 2013 – http://europa.eu/rapid/press-release_DOC-13-4_en.htm

——— (2014), *Conclusions on 2030 Climate and Energy Policy Framework, EUCO 169/14, 23 and 24 October 2014*, http://www.consilium.europa.eu/uedocs/cms_data/docs/pressdata/en/ec/145397.pdf [1 December 2014].

——— (2015), *European Council Conclusions, 19–20 March 2015, EUCO 11/15*, http://www.consilium.europa.eu/en/meetings/european-council/2015/03/19-20/ [4 May 2015]

European Energy Exchange (EEX). (2014), *Strong Growth on the EEX Markets in 2013*, Leipzig: European Energy Exchange AG, http://www.eex.com/blob/66776/8ea17d8bc1445c4113d8774772d6dee8/pr-20140114-eex-jahresrueckblick-pdf-data.pdf [3 October 2014].

European Environment Agency. (2008), *Greenhouse Gas Emission Trends and Projections in Europe 2008. Tracking Progress Ttowards Kyoto Targets*, http://www.eea.europa.eu/publications/eea_report_2008_5/Trends_and_projections_2008_executive_summary.pdf [30 April 2014].

——— (2013), *Final Energy Consumption by Sector (CSI 027/ENER 016) – Assessment Published February 2013*, http://www.eea.europa.eu/

data-and-maps/indicators/final-energy-consumption-by-sector-5/assessment [30 March 2014].

——— (2013a), *Transport Emissions of Greenhouse Gases (TERM 002) – Assessment Published December 2013*, http://www.eea.europa.eu/data-and-maps/indicators/transport-emissions-of-greenhouse-gases/transport-emissions-of-greenhouse-gases-3 [30 April 2014].

European External Action Service (EEAS). (2009), *EU-US Summit Declaration of 3 November 2009*, http://eeas.europa.eu/us/sum11_09/docs/declaration_en.pdf [30 April 2015].

——— (2009a), *The EU-US Energy Council (Annex 2)*, http://eeas.europa.eu/us/sum11_09/docs/energy_en.pdf [30 March 2015].

European Parliament. (1994), *Resolution on 1994 being Declared Community Energy Year* [OJ 1991, C 267/201].

——— (1999), *Rules Governing the Establishment of Intergroups*, http://www.europarl.europa.eu/pdf/intergroupes/legislature_8/DecisionofCoPof16Dec1999-Rules_governing_the_establishment_of_integroupes.pdf [30 April 2015].

——— (2003), *Lobbying in the European Union: Current Rules and Practices, Constitutional Affairs Series AFCO 104 EN*, http://ec.europa.eu/civil_society/interest_groups/docs/workingdocparl.pdf [30 April 2014].

——— (2005), *Draft Recommendation for Second Reading on the Council Common Position for Adopting a Directive of the European Parliament and of the Council on Establishing a Framework for the Setting of Eco-Design Requirements for Energy-Using Products and Amending Council Directive 92/42/EEC and Directives 96/57/EC and 2000/55/EC of the European Parliament and of the Council (11414/1/2004–C6-0246/2004 –2003/0172(COD))*, http://www.europarl.europa.eu/meetdocs/2004_2009/documents/pr/551/551338/551338en.pdf [30 April 2015].

——— (2005a), *European Parliament Legislative Observatory, Ecodesign Requirements for Energy-Using Products*, http://www.europarl.europa.eu/oeil/popups/ficheprocedure.do?lang=en&reference=2003/0172%28COD%29 [30 April 2015].

——— (2008), *Report on '2050: The Future Begins Today – Recommendations for the EU's Future Integrated Policy on Climate Change'* (2008/2105(INI), 10 December 2008), Temporary Committee on Climate Change, http://www.europarl.europa.eu/sides/getDoc.do?type=REPORT&reference=A6-2008-0495&language=EN [30 April 2014].

——— (2014), *Rules of Procedure, 7th Parliamentary Term, Strasbourg*, http://www.europarl.europa.eu/sides/getDoc.do?pubRef=-//EP//NONSGML+RULES-EP+20100705+0+DOC+PDF+V0//EN&language=EN [30 April 2014].

——— (2014a), *2030 Climate Goals: MEPs Sceptical about Commission Proposals, Press Release, Strasbourg*, http://www.europarl.europa.eu/pdfs/news/expert/infopress/20140121IPR33311/20140121IPR33311_en.pdf [30 April 2015].

——— (2015), *European Parliament Legislative Observatory Procedure File 2011/0092(CNS), Taxation of Energy Products and Electricity: Restructuring the Community Framework*, http://www.europarl.europa.eu/oeil/popups/ficheprocedure.do?reference=2011/0092%28CNS%29&l=en#keyEvents [30 April 2015].

European Parliament and Council of the European Union. (1996), *Directive 96/92/EC of 19 December 1996 Concerning Common Rules for the Internal Market in Electricity* [OJ 1997, L 27/20].

——— (1998), *Directive 98/30/EC of the European Parliament and of the Council of 22 June 1998 Concerning Rules for the Internal Market in Natural Gas* [OJ 1998, L 204/1].

——— (2001), *Directive 2001/77/EC of the European Parliament and of the Council of 27 September 2001 on the Promotion of Electricity from Renewable Energy Sources in the Internal Electricity Market* [OJ 2001, L 283/33].

——— (2002), *Directive 2002/91/EC of the European Parliament and of the Council of 16 December 2002 on the Energy Performance of Buildings* [OJ 2003, L 1/65].

——— (2003), *Regulation 1228/2003 of the European Parliament and of the Council of 26 June 2003 on Conditions for Access to the Network for Cross-Border Exchanges in Electricity* [OJ 2003, L176/1].

——— (2003a), *Directive 2003/30/EC of the European Parliament and of the Council of 8 May 2003 on the Promotion of the Use of Biofuels or Other Renewable Fuels for Transport* [OJ 2003, L 123/42].

——— (2003b), *Directive 2003/54/EC of the European Parliament and of the Council of 26 June 2003 Concerning Common Rules for the Internal Market in Electricity and Repealing Directive 96/92/EC* [OJ 2003, L 176/37].

——— (2003c), *Directive 2003/55/EC of the European Parliament and of the Council of 26 June 2003 Concerning Common Rules for the Internal Market in Natural Gas and Repealing Directive 98/30/EC* [OJ 2003, L176/57].

——— (2003d), *Directive 2003/87/EC of the European Parliament and of the Council of 13 October 2003 Establishing a Scheme for Greenhouse Gas Emission Allowance Trading within the Community and Amending Council Directive 96/61/EC* [OJ 2003, L 275/32].

——— (2004), *Directive 2004/101/EC of the European Parliament and of the Council of 27 October 2004 Amending Directive 2003/87/EC Establishing a Scheme for Greenhouse Gas Emission Allowance Trading within the Community, in Respect of the Kyoto Protocol's Project Mechanisms* [OJ 2004, L 338/18].

——— (2004a), *Directive 2004/67/EC of the European Parliament and of the Council of 26 April 2004 Concerning Measures to Safeguard Security of Natural Gas Supply* [OJ 2004, L 127/92].

——— (2005), *Directive 2005/89/EC of the European Parliament and of the Council of 18 January 2006 Concerning Measures to Safeguard*

Security of Electricity Supply and Infrastructure Investment [OJ 2006, L 33/22].

——— (2005a), *Directive 2005/32/EC of the European Parliament and of the Council of 6 July 2005 Establishing a Framework for the Setting of Ecodesign Requirements for Energy-Using Products and Amending Council Directive 92/42/EEC and Directives 96/57/EC and 2000/55/EC of the European Parliament and of the Council* [OJ 2005, L 191/29].

——— (2008), *Directive 2008/101/EC of the European Parliament and of the Council of 19 November 2008 Amending Directive 2003/87/ EC so as to Include Aviation Activities in the Scheme for Greenhouse Gas Emission Allowance Trading within the Community* [OJ 2009, L 8/3].

——— (2008a), *Directive 2008/68/EC of the European Parliament and of the Council of 24 September 2008 on the Inland Transport of Dangerous Goods* [OJ 2008, L 260/13].

——— (2009), *Decision 406/2009/EC of the European Parliament and of the Council of 23 April 2009 on the Effort of Member States to Reduce their Greenhouse Gas Emissions to Meet the Community's Greenhouse Gas Emission Reduction Commitments Up to 2020* [OJ 2009, L 140/136].

——— (2009a), *Directive 2009/72/EC of the European Parliament and of the Council of 13 July 2009 Concerning Common Rules for the Internal Market in Electricity and Repealing Directive 2003/54/EC* [OJ 2009, L 211/55].

——— (2009b), *Directive 2009/73/EC of the European Parliament and of the Council of 13 July 2009 Concerning Common Rules for the Internal Market in Natural Gas and Repealing Directive 2003/55/EC* [OJ 2009, L 211/94].

——— (2009c), *Directive 2009/29/EC of the European Parliament and of the Council of 23 April 2009 Amending Directive 2003/87/EC so as to Improve and Extend the Greenhouse Gas Emission Allowance Trading Scheme of the Community* [OJ 2009, L 140/63].

——— (2009d), *Directive 2009/28/EC of the European Parliament and of the Council of 23 April 2009 on the Promotion of the Use of Energy from Renewable Sources and Amending and Subsequently Repealing Directives 2001/77/EC and 2003/30/EC* [OJ 2009, L 140/16].

——— (2009e), *Regulation (EC) No. 443/2009 of the European Parliament and of the Council of 23 April 2009 Setting Emission Performance Standards for New Passenger Cars as Part of the Community's Integrated Approach to Reduce CO_2 Emissions from Light-Duty Vehicles* [OJ 2009, L 140/1]

——— (2009f), *Regulation (EC) No. 713/2009 of the European Parliament and of the Council of 13 July 2009 Establishing an Agency for the Cooperation of Energy Regulators* [OJ 2009, L 211/1].

——— (2009g), *Regulation (EC) No. 714/2009 of the European Parliament and of the Council of 13 July 2009 on Conditions for Access to the*

Network for Cross-Border Exchanges in Electricity and Repealing Regulation (EC) No. 1228/2003 [OJ 2009, L 211/15].

——— (2009h), *Regulation (EC) No. 715/2009 of the European Parliament and of the Council of 13 July 2009 on Conditions for Access to the Natural Gas Transmission Networks and Repealing Regulation (EC) No. 1775/2005* [OJ 2009 L211/36].

——— (2009i), *Directive 2009/31/EC of the European Parliament and of the Council of 23 April 2009 on the Geological Storage of Carbon Dioxide and Amending Council Directive 85/337/EEC, European Parliament and Council Directives 2000/60/EC, 2001/80/EC, 2004/35/EC, 2006/12/EC, 2008/1/EC and Regulation (EC) No. 1013/2006* [OJ 2009, L140/114].

——— (2009j), *Directive 2009/125/EC of the European Parliament and of the Council of 21 October 2009 Establishing a Framework for the Setting of Ecodesign Requirements for Energy-Related Products* [OJ 2009, L 285/10].

——— (2010), *Directive 2010/30/EU of the European Parliament and of the Council of 19 May 2010 on the Indication by Labelling and Standard Product Information of the Consumption of Energy and Other Resources by Energy-Related Products* [OJ 2010, L 153/1].

——— (2010a), *Regulation (EU) No. 994/2010 of the European Parliament and of the Council of 20 October 2010 Concerning Measures to Safeguard Security of Gas Supply and Repealing Council Directive 2004/67/EC* [OJ 2010, L 295/1].

——— (2010b), *Council Regulation (EU, Euratom) No. 617/2010 of 24 June 2010 Concerning the Notification to the Commission of Investment Projects in Energy Infrastructure within the European Union and Repealing Regulation (EC) No. 736/96* [OJ 2010, L 180/7].

——— (2010c), *Decision No. 661/2010/EU of the European Parliament and of the Council of 7 July 2010 on Union Guidelines for the Development of the Trans-European Transport Network* [OJ 2010, L 204/1].

——— (2011), *Regulation (EC) No. 1227/2011 of the European Parliament and of the Council on 25 October 2011 on Wholesale Energy Market Integrity and Transparency* [OJ 2011, L 326/1].

——— (2012), *Directive of the European Parliament and of the Council of 25 October 2012 on Energy Efficiency, Amending Directives 2009/125/EC and 2010/30/EC and Repealing Directives 2004/8/EC and 2006/32/EC*, 2012/27/EU [OJ 2012, L 315/1].

——— (2013), *Directive of the European Parliament and of the Council of 12 June 2013 on Safety of Offshore Oil and Gas Operations and Amending Directive 2004/35/EC*, 2013/30/EU [OJ 2013, L 178/66].

——— (2013a), *Regulation (EU) No. 347/2013 of the European Parliament and of the Council of 17 April 2013 on Guidelines for Trans-European Energy Infrastructure and Repealing Decision No. 1364/2006/EC and Amending Regulations (EC) No. 713/2009, (EC) No. 714/2009 and (EC) No. 715/2009* [OJ 2013, L 115/39].

——— (2014), *Regulation (EU) No. 256/2014 of the European Parliament and of the Council of 26 February 2014 Concerning the Notification to the Commission of Investment Projects in Energy Infrastructure within the European Union, Replacing Council Regulation (EU, Euratom) No. 617/2010 and Repealing Council Regulation (EC) No. 736/96* [OJ 2014, L 084/61].

European Parliament and European Commission. (2011), *Agreement between the European Parliament and the European Commission on the Establishment of a Transparency Register for Organisations and Self-Employed Individuals Engaged in EU Policymaking and Policy Implementation* [OJ 2001, L 191/29].

——— (2014), *Agreement Between the European Parliament and the European Commission on the Transparency Register for Organisations and Self-Employed Individuals Engaged in EU Policy-making and Policy Implementation* [OJ 2014, L 277/11].

——— (2015), *Statistics for the Transparency Register*, http://ec.europa.eu/transparencyregister/public/consultation/statistics.do?locale=en&action=prepareView [30 April 2015].

European Wind Energy Association (EWEA). (2009), *Pure Power. Wind Energy Targets for 2020 and 2030. A Report by the European Wind Energy Association-2009 Update*, Brussels, http://www.ewea.org/fileadmin/files/library/publications/reports/Pure_Power_2009.pdf [30 April 2014].

——— (2014), *The European Offshore Wind Industry – Key Trends and Statistics 1st Half 2014*, Brussels, http://www.ewea.org/fileadmin/files/library/publications/statistics/European_offshore_statistics_1st-half_2014.pdf [30 April 2015].

Eurostat. (2007), *Gas and Electricity Market Statistics 2007 Edition*, http://ec.europa.eu/eurostat/documents/3217494/5620400/KS-GB-07-001-EN.PDF/24cf251a-7126-4c50-8b9d-9ae8ab229e55?version=1.0 [30 April 2014].

——— (2009), *Air Transport Sector Statistics – NACE Rev. 1.1*, http://epp.eurostat.ec.europa.eu/statistics_explained/index.php/Air_transport_sector_statistics_-_NACE_Rev._1.1 [23 September 2014].

——— (2012), *Main Origin of Primary Energy Imports, EU-27, 2002–2010 (% of Extra EU-27 Imports)*, http://epp.eurostat.ec.europa.eu/statistics_explained/index.php/Energy_production_and_imports [17 September 2013].

——— (2012a), *Share of Renewables in Gross Final Energy Consumption, 2010 and 2020 (%)*, http://epp.eurostat.ec.europa.eu/statistics_explained/index.php?title=File:Share_of_renewables_in_gross_final_energy_consumption,_2010_and_2020_%28%25%29.png&filetimestamp=20121012133709 [9 October 2013].

——— (2012c), *Imports – Oil – Annual Data*, http://appsso.eurostat.ec.europa.eu/nui/show.do?dataset=nrg_123a&lang=en [20 May 2012].

——— (2013), *Energy, Transport and Environment Indicators,* Eurostat Pocketbooks, Luxembourg: Publications Office of the European Union.

——— (2014), *Share of Renewable Energy in Fuel Consumption of Transport – %*, http://epp.eurostat.ec.europa.eu/tgm/table.do?tab=table&init=1&plugin=1&language=en&pcode=tsdcc340 [30 April 2014].

——— (2014a), *Share of Energy from Renewable Sources*, http://appsso.eurostat.ec.europa.eu/nui/show.do?dataset=nrg_ind_335a&lang=en [20 May 2014].

——— (2014b), *Supply, Transformation, Consumption – Electricity – Annual Data*, http://appsso.eurostat.ec.europa.eu/nui/show.do?dataset=nrg_105a&lang=en [20 May 2014].

——— (2014c), *Supply, Transformation, Consumption – Renewable Energies – Annual Data*, http://appsso.eurostat.ec.europa.eu/nui/show.do?dataset=nrg_107a&lang=en [20 May 2014].

——— (2014d), *Electricity Production and Supply Statistics,* http://epp.eurostat.ec.europa.eu/statistics_explained/index.php/Electricity_production_and_supply_statistics [20 May 2014].

——— (2014e), *Supply – Electricity – Monthly Data,* http://appsso.eurostat.ec.europa.eu/nui/show.do?dataset=nrg_105m&lang=en [20 May 2014].

——— (2014f), *Supply, Transformation, Consumption – All Products – Annual Data*, http://appsso.eurostat.ec.europa.eu/nui/show.do?dataset=nrg_100a&lang=en [20 May 2014].

——— (2014g), *GDP and Main Components – Volumes,* http://appsso.eurostat.ec.europa.eu/nui/show.do?dataset=nama_gdp_k&lang=en [20 May 2014].

——— (2014h), *Supply Transformation and Consumption of Electricity – Annual Data*, http://appsso.eurostat.ec.europa.eu/nui/show.do?dataset=nrg_105a&lang=en [30 March 2014].

——— (2014i), *Combined Heat and Power Generation*, http://epp.eurostat.ec.europa.eu/tgm/table.do?tab=table&init=1&plugin=1&language=en&pcode=tsdcc350 [30 April 2014].

——— (2014j), *Modal Split of Passenger Transport*, http://appsso.eurostat.ec.europa.eu/nui/show.do?dataset=tran_hv_psmod&lang=en [30 April 2014].

——— (2014k), *Air Passenger Transport in the EU27*, http://epp.eurostat.ec.europa.eu/cache/ITY_PUBLIC/7-05112013-BP/EN/7-05112013-BP-EN.PDF [30 April 2014]. See also: http://ec.europa.eu/eurostat/statistics-explained/index.php/Passenger_transport_statistics [30 April 2015].

——— (2014l), *Final Consumption Expenditure of Households, by Consumption Purpose*, http://epp.eurostat.ec.europa.eu/tgm/table.do?tab=table&init=1&plugin=1&language=en&pcode=tsdpc520 [30 April 2014].

——— (2014m), *Simplified Energy Balances – Annual Data,* http://appsso.eurostat.ec.europa.eu/nui/show.do?dataset=nrg_100a&lang=en [20 May 2014].

——— (2014n), *Final Energy Consumption by Sector*, http://epp.eurostat.ec.europa.eu/tgm/table.do?tab=table&init=1&language=en&pcode=tsdpc320&plugin=1 [20 May 2012].

——— (2014o), *Electricity Consumption by Households,* http://ec.europa.eu/eurostat/tgm/table.do?tab=table&init=1&language=en&pcode=tsdpc310&plugin=1 [30 April 2014].

——— (2014p), *Electricity Production, Consumption and Market Overview,* http://ec.europa.eu/eurostat/statistics-explained/index.php/Electricity_production,_consumption_and_market_overview [30 April 2014].

——— (2014q), *Electricity Prices for Domestic Consumers – Bi-annual Data (from 2007 onwards),* http://appsso.eurostat.ec.europa.eu/nui/show.do?dataset=nrg_pc_204&lang=en [30 March 2014].

——— (2014r), *Gas Prices for Domestic Consumers – Bi-annual Data (from 2007 onwards),* http://appsso.eurostat.ec.europa.eu/nui/show.do?dataset=nrg_pc_202&lang=en [30 April 2014].

——— (2015), *Energy Dependence,* http://epp.eurostat.ec.europa.eu/tgm/refreshTableAction.do?tab=table&plugin=1&pcode=tsdcc310&language=en [30 April 2015].

——— (2015a), *Energy Dependence*, http://epp.eurostat.ec.europa.eu/tgm/mapToolClosed.do?tab=map&init=1&plugin=1&language=en&pcode=tsdcc310&toolbox=legend [30 April 2015].

——— (2015b), *Exports – Electricity – Annual Data,* http://appsso.eurostat.ec.europa.eu/nui/show.do?dataset=nrg_135a&lang=en [30 April 2015].

——— (2015c), *Imports – Electricity – Annual Data*, http://appsso.eurostat.ec.europa.eu/nui/show.do?dataset=nrg_125a&lang=en [30 April 2015].

——— (2015d), *Imports – Gas – Annual Data*, http://appsso.eurostat.ec.europa.eu/nui/show.do?dataset=nrg_124a&lang=en [30 April 2015].

——— (2015e), *Agri-Environmental Indicator – Renewable Energy Production*,http://ec.europa.eu/eurostat/statistics-explained/index.php?title=Agri-environmental_indicator_-_renewable_energy_production [30 April 2015].

——— (2015f), *Greenhouse Gas Emissions,* http://ec.europa.eu/eurostat/tgm/table.do?tab=table&plugin=1&language=en&pcode=tsdcc100 [30 April 2015].

——— (2015g), *Share of Renewable Energy in Gross Final Energy Consumption*, http://ec.europa.eu/eurostat/tgm/table.do?tab=table&init=1&language=en&pcode=t2020_31&plugin=1 [30 April 2015].

——— (2015h), *Renewable Energy in the EU, Share of Renewables in Energy Consumption up to 15% in the EU in 2013*, Eurostat News Release, 43/2015, 10 March 2015, http://ec.europa.eu/eurostat/documents/2995521/6734513/8-10032015-AP-EN.pdf [30 April 2015].

Fernandez Martin, R.M. (2012), 'The European Union and International Negotiations on Climate Change: A Limited Role to Play', *Journal of Contemporary European Research*, 8(2), 193–209.

Ferrier, J. (2013), *Reconciling Market Trading with Long-term Partnerships, Presentation to the RGS-Eurogas Conference, Energy Dialogue: Russia-EU. Gas Aspects (30 May 2013),* http://www.eurogas.org/uploads/media/RGS-Eurogas_Conference_30_May_2013_SESSION_2_Ferrier.pdf [30 April 2014].

Florence Forum. (2015), *28th Meeting of the European Electricity Regulatory Forum, Florence, 4 June 2015 to 5 June 2015, Convitto della Calza, Florence,* http://ec.europa.eu/energy/en/events/28th-meeting-european-electricity-regulatory-forum-florence [30 April 2015].

Franchino, F. (2007), *The Powers of the Union. Delegation in the EU,* Cambridge: University Press.

Frank, R.H. (2006), 'Energy Policy is Far too Complicated to be Left to the Politicians' [8 June 2006], *New York Times,* http://www.nytimes.com/2006/06/08/business/08scene.html?_r=2&oref=slogin&pagewanted=print [30 April 2014].

Freestone, D. and Streck, C. (Eds.). (2005), *Legal Aspects of Implementing the Kyoto Protocol Mechanism: Making Kyoto Work*, Oxford: Oxford University Press.

GAC. (2013), *Conclusions of the 8th EU-Russia Gas Advisory Council Moscow, 19 November 2013*, Moscow: Gas Advisory Council, http://ec.europa.eu/energy/sites/ener/files/documents/20131119_gac_conclusions.pdf [30 April 2015].

Gas LNG Europe. (2015), *Aggregated LNG Storage Inventory,* http://lngdataplatform.gie.eu/index.php [30 April 2015].

Gassco. (2014), *Norwegian Natural Gas Transport System,* http://www.gassco.no/static/transport/ [30 April 2014]

Gazprom. (2014), *Gazprom in Questions and Answers*, Moscow: OAO Gazprom, http://www.gazpromquestions.ru/en/ [30 April 2015].

Geden, O. and Fischer, S. (2008), *Die Energie- und Klimapolitik der Europäischen Union. Bestandaufnahme und Perspektiven*, Baden-Baden: Nomos.

Glachant, J.M., Grant, R., Hafner, M. and de Jong, J. (2010), *Toward a Smart EU Energy Policy: Rationale and 22 Recommendations*, Policy Brief of The Florence School of Regulation (FSR), European University Institute, ISSUE 2010/03.

Glover, P.C. and Economides, M.J. (2010), *Energy and Climate Wars*, New York: The Continuum International Publishing Group.

Gloystein, H. and Kahn, M. (2014), 'Europe Less Reliant on Russian Gas through Ukraine' [3 March 2014], *Reuters US Edition*, London: Reuters, http://www.reuters.com/article/2014/03/03/ukraine-crisis-gas-europe-idUSL6N0M01VY20140303 [30 April 2014].

Goldstein, H. and Sweet, W. (2007), 'Joules, BTUs, Quads—Let's Call the Whole Thing Off' [1 January 2007], *IEEE Spectrum Online*, http://www.spectrum.ieee.org/jan07/4820 [30 April 2014].

Graphene Flagship. (2015), *Graphene Flagship*, http://graphene-flagship.eu [30 April 2015].

Green 10. (2005), *Letter to the European Commission Regarding its Forthcoming Communication Reducing the Climate Change Impact of Aviation* [11 March 2014].

——— (2007), *Reducing CO_2 Emissions from Light Duty Vehicles* [23 January 2007].

——— (2009), *Green Groups: European Commission off Target* [10 June 2009].

Groenendijk. (2014), *Use and Role of LNG in 2013 and Expected Contribution to SoS in 2014, Presentation to the Gas Coordination Group, Brussels, 14 February 2014*, http://www.gie.eu/index.php/publications/doc_download/21681-gle-presentation-to-the-gcg-use-and-role-of-lng-in-2013-and-expected-contribution-to-sos-in-2014 [30 April 2015].

Grunwald, J. (2003), *Das Energierecht der Europäischen Gemeinschaften. EGKS – Euratom – EG. Grundlagen – Geschichte – Geltende Regelungen*, Berlin: de Gruyter.

Gulmeza, S.B. (2013), 'Rising Euroscepticism in Turkish Politics: The Cases of the AKP and the CHP', *Acta Politica*, 48, 326–344.

Haase, N. (2009), *European Gas Market Liberalization*, Groningen: Energy Delta Institute.

Hedegaard, C. (2014), *Setting the Course for Low-Carbon Growth*, http://ec.europa.eu/commission_2010-2014/hedegaard/headlines/articles/2014-02-07_01_en.htm [30 April 2014].

Helm, D. (Ed.). (2005), *Climate-Change Policy*, Oxford: Oxford University Press.

——— (2012), *The Carbon Crunch. How We're Getting Climate Change Wrong – and How to Fix It*, New Haven: Yale University Press.

——— (2012a), 'To Slow Warming, Tax Carbon', *New York Times*, 11 November 2012, A29.

——— (2013), *European Energy Policy in the New Geopolitical Context, Presentation made at the Clubroom of the Foreign Policy And United Nations Association of Austria*, http://www.dieterhelm.co.uk/sites/default/files/International%20Club%20231013%20FINAL.pdf [1 August 2015].

Helm, D. and Hepburn, C. (2010), *The Economics and Politics of Climate Change*, Oxford: Oxford University Press.

Hofer, S. (2008), *Die Europäische Union als Regelexporteur. Die Europäisierung der Energiepolitik in Bulgarien, Serbien und der Ukraine*, Baden-Baden: Nomos.

Hoffmann, S. (1966), 'Obstinate or Obsolete? The Fate of the Nation-State and the Case of Western Europe', *Daedalus*, 3, 862–915.

——— (1982), 'Reflections on the Nation-State in Western Europe Today', *Journal of Common Market Studies*, 21(1), 21–37.

Hofstrand, D. (2008), *Natural Gas and Coal Measurements and Conversions, File C6-89*, Iowa State University, https://www.extension.iastate.edu/agdm/wholefarm/html/c6-89.html [30 April 2014].

Hopkirk, P. (1994), *The Great Game: The Struggle for Empire in Central Asia,* New York: Kodansha America.

Housiadas, C. and Dimitriou, P. (2013), *The Current Radiation and Nuclear Safety Regulatory Framework and its Revision at EU Level, Presentation to the Workshop on the Revision of the European Regulatory Framework in the Fields of Radiological and Nuclear Safety – The Regulators' Approach in the Light of Four Successive Presidencies of the Council of the European Union, Crowne Plaza Hotel, Athens, Monday*, http://eeae.gr/docs/files/_housiadas_dimitriou.pdf [1 August 2015].

Howarth, R.W., Bringezu, S., Bekunda, M., de Fraiture, C., Maene, L., Martinelli, L. and Sala, O. (2009), 'Rapid assessment on biofuels and the environment: Overview and key findings', in: Howarth, R.W. and Bringezu, S. (Eds.), *Biofuels: Environmental Consequences*, Ithaca: Cornell University, pp. 1–13.

Howes, T. (2010), 'The EU's New Renewable Energy Directive (2009/28/EC)', in: Oberthür, S. and Pallemaerts, M. (Eds.), *The New Climate Change Policies of the European Union*, Brussels: VUB, pp. 117–150.

Ibrahimov, R. (2007), 'Prospects of The Russian Oil Pipeline to The West, Advantages and Disadvantages', *The Journal of the Turkish Weekly Opinion, USAK Energy Review of Current Infrastructure*, 8, 8–10.

International Atomic Energy Agency (IAEA). (2006), *The Chernobyl Forum: 2003-2005. Chernobyl's Legacy: Health, Environmental and Socio-Economic Impacts and Recommendations to the Governments of Belarus, the Russian Federation and Ukraine*, Vienna: IAEA.

IGU. (2012), *Natural Gas Conversion Pocketbook*, Fornebu, Norway: International Gas Union.

——— (2013), *World LNG Report – 2013 Edition*, Fornebu, Norway: International Gas Union.

International Air Transport Association. (2014), *Operational Fuel Efficiency*, http://www.iata.org/whatwedo/ops-infra/Pages/fuel-efficiency.aspx [30 April 2014].

International Energy Agency. (1997), *Oil in Power Generation,* Paris: OECD/IEA.

——— (1998), *Caspian Oil and Gas: The Supply Potential of Central Asia and Transcaucasia/International Energy Agency, in Co-operation with the Energy Charter Secretariat,* Paris: OECD/IEA.

——— (2008), *From 1st- to 2nd-Generation Biofuel Technologies. An Overview of Current Industry and RD&D Activities*, Paris: OECD/IEA.

——— (2012), *CO_2 Emissions from Fuel Combustion - Highlight*, Paris: IEA.

——— (2012a), *World Energy Outlook 2012*, Paris: OECD/IEA.
——— (2013), *World Energy Outlook 2013*, Paris: OECD/IEA.
——— (2013a), *Key World Energy Statistics*, Paris: OECD/IEA.
——— (2015), *Glossary*, http://www.iea.org/aboutus/glossary/ [30 April 2015].
IPCC. (1999), *Aviation and the Global Atmosphere – Summary for Policy-makers*, http://www.grida.no/publications/other/ipcc_sr/ [30 April 2014].
——— (2007), *IPCC Fourth Assessment Report – Climate Change 2007. Synthesis Report*, http://www.ipcc.ch/pdf/assessment-report/ar4/syr/ar4_syr.pdf [30 April 2014].
Isachenkov, V. (2003), 'China's Hu Calls for Multipolar World, Russian and Chinese Companies Strike Oil Deal' [28 May 2003], *Associated Press Worldstream*, http://www.highbeam.com/doc/1P1-74198449.html [1 August 2015].
Jasper, U. and Portela, C. (2010), 'EU Defence Integration and Nuclear Weapons: A Common Deterrent for Europe?' *Security Dialogue*, 41(2), 145–168.
Jevons, W.S. (2008) [1866], *The Coal Question: An Inquiry Concerning the Progress of the Nation, and the Probable Exhaustion of Our Coal Mines*, London: Dodo Press.
Juncker, J.C. (2014), *A New Start for Europe: My Agenda for Jobs, Growth, Fairness and Democratic Change, Opening Statement in the European Parliament Plenary Session, Strasbourg, 15 July 2014*, http://ec.europa.eu/priorities/docs/pg_en.pdf#page=6 [20 January 2015].
Kamm, H. (1968), 'Cernik Meets with Kremlin Leaders in Day's Visit', *New York Times,* 11 November 1968, 3.
Karbuz, S. (2010), 'Losing the Energy Battle: How and Why the US and EU Need to Engage the Black Sea Region', *Journal of Energy Security*, 10 July 2010, http://ensec.org/index.php?option=com_content&view=article&id=255:losing-the-battle-why-and-how-the-united-states-and-europe-need-to-engage-the-black-sea-region&catid=108:energysecuritycontent&Itemid=365 [1 August 2015].
Kearns, M., Suri, S. and Montfort, N. (2006), 'An Experimental Study of the Coloring Problem on Human Subject Networks', *Science*, 313(5788), 824–827.
Konstatyan, H. (2012), *The EEAS and the Eastern Partnership: Let the Blame Game Stop*, *CEPS Commentary,* Brussels, http://www.ceps.be/book/eeas-and-eastern-partnership-let-blame-game-stop [20 May 2014].
Koranyi, D. (Ed.). (2011), *Transatlantic Energy Futures. Strategic Perspectives on Energy Security, Climate Change and New Technologies in Europe and the United States*, Washington, DC: Center for Transatlantic Relations.
KPMG. (2012), *Central and Eastern European Shale Gas Outlook*, London, https://www.kpmg.com/Global/en/IssuesAndInsights/ArticlesPublications/shale-gas/Documents/cee-shale-gas-1.pdf [20 October 2014].

Kreutler, M. (2014), *Interest Group Coalitions in the European Union: An Analysis of (In)Formal Initiatives to Influence European Energy Policy,* Baden-Baden: Nomos.

Kroes, N. (2007), *Introductory Remarks on Final Report of Energy Sector Inquiry*, Speech/07/4, Brussels, 10 January.

Lesage, D., van de Graaf, T. and Westphal, K. (2010), *Global Energy Governance in a Multipolar World*, Farnham: Ashgate.

LNG Blue Corridors. (2015), *LNG Blue Corridors: Demonstration of Heavy-Duty Vehicles Running with Liquefied Methane*, http://lngbc.eu/ [30 April 2015].

Lyons, P.K. (1998), *EU Energy Policies towards the 21st Century*, EC INFORM.

Madrid Forum. (2013), *Conclusions (15–16 October 2013), Madrid*, http://www.ceer.eu/portal/page/portal/EER_HOME/EER_WORKSHOP/Stakeholder%20Fora/Madrid%20Fora/24th_Madrid_Forum/meeting_024_conclusions.pdf [30 April 2015].

Matláry, J.H. (1997), *Energy Policy in the European Union*, New York: St. Martin's Press.

MEES. (2008), 'Caspian Sea Region Pipelines – Existing and Proposed', *Middle East Economic Survey,* LI(11), 28.

Milward, A. (1984), *The Reconstruction of Western Europe 1945–1951*, London: Methuen.

——— (1994), *The European Rescue of the Nation State,* London: Routledge.

Moravcsik, A. (1993), 'Preferences and Power in the European Community. A Liberal Intergovernmentalist Approach', *Journal of Common Market Studies*, 31(4), 473–524.

——— (1998), *The Choice for Europe. Social Purpose and State Power from Messina to Maastricht,* Ithaca: Cornell University Press.

Morelli, V.L. (2013), *European Union Enlargement: A Status Report on Turkey's Accession Negotiations (RS22517)*, Washington, DC: Congressional Research Service.

Norwegian Petroleum Directorate. (2008), *FACTS 2008 The Norwegian Petroleum Sector 2008*, Stavanger: Ministry of Petroleum and Energy.

——— (2012), *FACTS 2012 The Norwegian Petroleum Sector 2012*, Stavanger: Ministry of Petroleum and Energy.

——— (2013), *FACTS 2013 The Norwegian Petroleum Sector 2013*, Stavanger: Ministry of Petroleum and Energy.

——— (2014), *FACTS 2014 The Norwegian Petroleum Sector 2014*, Stavanger: Ministry of Petroleum and Energy.

Novoselov, K.S., Geim, A.K., Morozov, S.V., Jiang, D., Zhang, Y., Dubonos, S.V., Grigorieva, I.V. and Firsov, A.A. (2004), 'Electric Field Effect in Atomically Thin Carbon Films', *Science,* 306(5696), 666–669.

Nowak, B. (2010), 'Energy Market of the European Union: Common or Segmented?' *The Electricity Journal, 23*(10), 27–37.

Nuttall, W.J. (2010), 'Nuclear energy in an enlarged Europe', in: Lévêque, F., et al. (Eds.), *Security of Energy Supply in Europe: Natural Gas, Nuclear and Hydrogen*, Cheltenham: Edward Elgar, pp. 166–189.

Nuttall, W.J. and Newberry, D.M. (2010), 'European electricity supply security and nuclear power: An overview' in: Lévêque, F., et al. (Eds.), *Security of Energy Supply in Europe: Natural Gas, Nuclear and Hydrogen*, Cheltenham: Edward Elgar, pp. 109–116.

Oberthür, S. and Pallemaerts, M. (Eds.). (2010), 'The EU's internal and external climate policies: An historical overview', in: Oberthür, S. and Pallemaerts, M. (Eds.), *The New Climate Policies of the European Union: Internal Legislation and Climate Diplomacy*, Brussels: Brussels University Press, pp. 27–64.

OECD. (2014), *The Cost of Air Pollution: Health Impacts of Road Transport*, OECD Publishing, http://dx.doi.org/10.1787/9789264210448-en [20 May 2014].

OFID. (2009), *Biofuels and Food Security: Implications of an Accelerated Biofuels Production*, Vienna: OFID.

Oliker, O. and Schlapak, D. (2005), *U.S. Interests in Central Asia: Policy Priorities and Military Roles*, Santa Monica: Rand Corporation.

Oliker, O., Crane, K., Schwartz, L.H. and Yusupov, C. (2009), *Russian Foreign Policy: Sources and Implications*. Santa Monica: RAND.

Österreich Institut für Wärme und Oeltechnik. (2012), *So heizt Österreich*, http://www.iwo-austria.at/fileadmin/user_upload/pdf_2012-1.HJ/pdf-2012-2._HJ/So_heizt_Oesterreich.pdf [30 March 2014].

Padgett, S. (1992), 'The Single European Energy Market: The Politics of Realization', *JCMS: Journal of Common Market Studies*, *30*(1), 53–76.

Pimentel, D. and Pimentel, M.H. (2007), *Food, Energy and Society*, Boca Raton, London, New York: CRC.

Platts. (2013), 'German Coal Extends Dominance in Power Mix as Gas Wanes' [9 October 2013], London: Platts, http://www.platts.com/latest-news/coal/london/analysis-german-coal-extends-dominance-in-power-26352497 [20 February 2014].

———(2014), *Energy Conversion Factors*, http://www.platts.com/conversion-tables [30 April 2014].

Pollack, M. (2010), 'Theorizing EU policy-making', in: Wallace, H., Pollack, M.A. and Young, A.R. (Eds.), *Policy-Making in the European Union,* Oxford: Oxford University Press, pp. 15–44.

Pollak, J. and Slominski, P. (2011), 'Liberalising the EU's energy market: Hard and soft power combined', in: Falkner, G. (Ed.), *The EU's Decision Traps: Comparing Policies*, Oxford: Oxford University Press, pp. 92–109.

Pollak, J., Schubert, S. and Slominski, P. (2010), *Die Energiepolitik der EU*, Wien: wuv.facultas.

Poynter, C. (2007), *European Gas & Power Markets' Quest for Supply Diversity and Security, Summer NAPE 2007*, Houston, TX: IHS Inc., p. 34.

Pratt, M. (2007), 'Lines in the Sea: Maritime Boundary Delimitation', *Hydro International,* 11(4), http://www.hydro-international.com/issues/articles/id772-Lines_in_the_Sea.html [30 April 2014].
PRC. (2006), *China's National Defense in 2006,* Beijing: Information Office of the State Council, PRC.
——— (2009), *China's National Defense in 2008,* Beijing: Information Office of the State Council.
Princen, S. and Kerremans, B. (2008), 'Opportunity Structures in the EU Multi-Level-System', *West European Politics*, 31(6), 1129–1146.
Prontera, A. (2009), 'Energy Policy: Concepts, Actors, Instruments and Recent Developments', *World Political Science Review*, 5(1), 1–29.
Putnam, R. (1988), 'Diplomacy and Domestic Politics: The Logic of Two-level Games', *International Organization*, 42(3), 427–460.
Ratner, M., Belken, P., Nichol, J. and Woehrel, S. (2013), *Europe's Energy Security: Options and Challenges to Natural Gas Supply Diversification (R42405)*, Washington, DC: Congressional Research Service.
Riley, A., Alex, B. and Rauscher, D. (2012), *Briefing Paper EU Energy Roadmap 2050: EU External Policies for Future Energy Security*, Brussels: European Union.
Scharpf, F.W. (1988), 'The Joint-Decision Trap: Lessons from German Federalism and European Integration', *Public Administration,* 66(3), 239–278.
Scharpf, F.W. (2006), 'The Joint-Decision Trap Revisited', *Journal of Common Market Studies*, 44(4), 845–864.
Scharples, J (2012), 'Russo-Polish Energy Security Relations: A Case of Threatening Dependency, Supply Guarantee, or Regional Energy Security Dynamics?' *Political Perspectives,* 6(1), 27–50.
Schmidt, R. (2003), 'Die Liberalisierung der Daseinsvorsorge', *Der Staat*, 42/2, 225–248.
Schmidt, S.K. (1996), 'Sterile Debates and Dubious Generalisations: European Integration Theory Tested by Telecommunications and Electricity', *Journal of Public Policy, 16*(3), 233–271.
——— (1998), *Liberalisierung in Europa. Die Rolle der Europäischen Kommission*, Frankfurt/Munich/New York: Campus.
——— (2000), 'Only an Agenda Setter?' *European Union Politics*, 1(1), 37–61.
——— (2001), 'Die Einflussmöglichkeiten der Europäischen Kommission auf die europäische Politik', *Politische Vierteljahresschrift*, 42(2), 173–192.
Schubert, S.R. (2006), 'Revisiting the Oil Curse', *Development*, 49(3), 64–70.
——— (2007), 'Being Rich in Energy Resources – A Blessing or a Curse', *EP Policy Department Studies*, 614(386), 1–60.
——— (2010), *Internal Substitution Policies and External Power Projection: How the Pursuit of Energy Autonomy Affects International Cooperation and Conflict: An Examination of the United States and the*

European Union, PhD Dissertation, Vienna: University of Vienna, http://othes.univie.ac.at/12955/ [30 April 2015].

Schulz, W. and Held, T. (2004), *Regulated Self-Regulation as a Form of Modern Government. An Analysis of Case Studies from Media and Telecommunications Law*, Eastleigh: University of Luton.

Schumann, D. (2003), 'Netzwerke und Koppelgeschäfte in der europäischen Elektrizitätspolitik', *Politische Vierteljahresschrift*, 44(3), 325–347.

Smale, A. (2014), 'On Ukraine, Merkel Finds Limits of Her Rapport with Putin' [12 March 2014], *New York Times Online*, http://www.nytimes.com/2014/03/13/world/europe/on-ukraine-merkel-finds-limits-of-her-rapport-with-putin.html?_r=0 [30 April 2014].

Smith, C. (2014), 'Product Pipeline Completions Lead Planned Construction Lower', *Oil & Gas Journal*, 2 March 2014.

Solana, J. (2006), *Address to the EU External Energy Policy Conference: Towards an EU External Energy Policy*, Brussels, 20 November, http://www.consilium.europa.eu/uedocs/cms_data/docs/pressdata/en/discours/91788.pdf [30 July 2015].

Special Council of Ministers. (1957), *Protokoll vom 8. Oktober 1957 (45. Tagung des Rates) zwischen Ministerrat und Hoher Behörde über Mittel und Wege zu einer koordinierten Energiepolitik*, Luxembourg [OJ 1957, A 35/574].

——— (1964), *Protokoll eines Abkommens betreffend die Energiefragen vereinbart zwischen den Regierungen der Mitgliedstaaten der Europäischen Gemeinschaft gelegentlich der 94. Tagung des Besonderen Ministerrats der Europäischen Gemeinschaft für Kohle und Stahl am 21. April 1964 in Luxemburg*, Luxembourg [OJ 1964, P 69/1099].

Spiegel. (2009), *The Battle for Gas. Europe's Pipeline War* [27 January 2009], http://www.spiegel.de/international/world/the-battle-for-gas-europe-s-pipeline-war-a-603803.html [30 April 2014].

Statements of the Foreign Ministers and other documents. (1973), *European Political Cooperation,* http://aei.pitt.edu/5576/1/5576.pdf [30 April 2014].

Statistisches Bundesamt. (2013), *Deutschland exportierte auch 2012 mehr Strom als es importierte*, https://www.destatis.de/DE/PresseService/Presse/Pressemitteilungen/2013/04/PD13_125_51.html [1 March 2014].

Stratfor. (2009), *Russia: South Stream Pipeline to be Commissioned in 2015* [10 March 2009], http://www.stratfor.com/situation-report/russia-south-stream-pipeline-be-commissioned-2015 [10 April 2014].

Strzelecki, M. (2011), *Shale-Gas Drilling Cost in Poland Triple U.S., Schlumberger Says*, Bloomberg, http://www.bloomberg.com/news/2011-11-29/shale-gas-drilling-cost-in-poland-triple-u-s-schlumberger-says.html [20 October 2014].

Sutter, R.G. (2012), *Chinese Foreign Relations: Power and Policy since the Cold War*, Plymouth: Rowman & Littlefield.

Sweeney, J.L. (2001), 'Energy Economics', in: Smelser, N.J. and Baltes, P.B. (Eds.), *International Encyclopedia of the Social and Behavioral Sciences*, Oxford: Elsevier Science, pp. 4513–4520.

Talseh, L.U. (2012), *The EU-Russia Energy Dialogue, Travelling without Moving*, Berlin: German Institute for International and Security Affairs.

ten Brink, P. (2010), 'Mitigating CO_2 emissions from cars in the EU (regulation (EC) No. 443/2009)', in: Oberthür, S. and Pallemaerts, M. (Eds.), *The New Climate Policies of the European Union. Internal Legislation and Climate Diplomacy*. Brussels: Brussels University Press, pp. 179–210.

Tsebelis, G. (2002), *Veto Players: How Political Institutions Work*, Princeton: Princeton University Press.

Tusk, D. (2014), 'A United Europe can End Russia's Energy Stranglehold', *Financial Times* [21 April 2014], http://www.ft.com/intl/cms/s/0/91508464-c661-11e3-ba0e-00144feabdc0.html#axzz3bd3zLzj1 [12 March 2015].

UNDP. (2008), *The Bali Action Plan: Key Issues in the Climate Negotiations – Summary for Policy Makers*, http://www.undp.org/content/dam/undp/library/Environment%20and%20Energy/Climate%20Change/Bali_Road_Map_Key_Issues_Under_Negotiation.pdf [30 August 2015].

UNFCCC. (1994), *The United Nations Framework Convention on Climate Change,* http://unfccc.int/essential_background/convention/items/2627.php [30 April 2014].

——— (1998), *Kyoto Protocol to The United Nations Framework Convention on Climate Change,* http://unfccc.int/resource/docs/convkp/kpeng.pdf [30 April 2015].

——— (2007), *Report of the Conference of the Parties on its Thirteenth Session, Held in Bali from 3 to 15 December 2007, Addendum Part Two: Action Taken by the Conference of the Parties at its Thirteenth Session,* http://unfccc.int/resource/docs/2007/cop13/eng/06a01.pdf [30 April 2015].

——— (2015), *First Steps to a Safer Future: Introducing The United Nations Framework Convention on Climate Change,* http://unfccc.int/essential_background/convention/items/6036.php [30 April 2015].

United Nations. (1982) *United Nations Convention on the Law of the Sea of 10 December 1982,* http://www.un.org/depts/los/convention_agreements/convention_overview_convention.htm [30 April 2014].

United Nations Scientific Committee on the Effects of Atomic Radiation (UNSCEAR). (2000), *Annex J, Exposures and Effects of the Chernobyl Accident,* http://www.unscear.org/docs/reports/2000/Volume%20II_Effects/AnnexJ_pages%20451-566.pdf [9 August 2015].

USGS. (2008), *Circum-Arctic Resource Appraisal: Estimates of Undiscovered Oil and Gas North of the Arctic Circle,* Washington, DC: USGS.

Van Vooren. (2012), *Europe Unplugged: Progress, Potential and Limitations of EU External Energy Policy Three Years Post-Lisbon*, Sieps Working Paper.

Vouldis, I., Millet, P. and Vallés, J.L. (2009), *Novel Materials for Energy Applications: A Decade of EU-funded Research,* Directorate-General for Research, Industrial Technologies, European Commission, http://ec.europa.eu/research/industrial_technologies/pdf/novel-materials-for-energy-applications_en.pdf [30 April 2015].

Weems, P.R. and Midani, F. (2009), *A Surprising Reality: Middle East Natural Gas Crunch*, *Who's Who Legal News,* http://www.whoswholegal.com/news/article/12773/a-surprising-reality-middle-east-natural-gas-crunch [30 April 2014].

Weiß, W. (2003), 'Öffentliche Unternehmen und EGV', *Europarecht,* 38(2), 165–190.

Wessels, W. (1991), 'The EC Council: The Community's decision-making center', in: Hoffmann, S. and Keohane, R.O. (Eds.), *The New European Community: Decision-Making and Institutional Change*, Boulder: Westview, pp. 133–154.

Westphal, K. (2011), 'Energy in an era of unprecedented uncertainty: International energy governance in the face of macroeconomic, geopolitical; and systemic challenges', in: Koranyi, D. (Ed.), *Transatlantic Energy Futures: Strategic Perspectives on Energy Security, Climate Change and New Technologies in Europe and the United States*, Washington: Center for Transatlantic Relations, pp. 1–26.

Wood, S. (2013), 'Turkey and Europe: Identity and Other Crises', *Turkish Studies*, 14(2), 272–291.

World Health Organization (WHO) Regional Office for Europe (2015). *Economic Cost of the Health Impact of Air Pollution in Europe: Clean Air, Health and Wealth*, Copenhagen: WHO Regional Office for Europe.

World Nuclear Association. (2014), *Uranium Production Figures, 2003–2013*, http://www.world-nuclear.org/info/Facts-and-Figures/Uranium-production-figures/ [30 April 2015].

World Trade Organization. (2014), *European Union and its Member States – Certain Measures Relating to The Energy Sector, Request for Consultations by The Russian Federation,* https://docs.wto.org/dol2fe/Pages/FE_Search/FE_S_S006.aspx?Query=(@Symbol=%20wt/ds476/*)&Language=ENGLISH&Context=FomerScriptedSearch&languageUIChanged=true# [2 October 2014].

Yergin, D. (2008), *The Prize: The Epic Quest for Oil, Money and Power*, New York: Free Press.

Youngs, R. (2011), 'Foreign policy and energy security: Markets, pipelines, and politics', in: Birchfield, V.L. and Duffield, J.S. (Eds.), *Toward a Common European Union Energy Policy. Problems, Progress, and Prospects*, New York: Palgrave Macmillan, pp. 41–60.

Index